Where?

Experiments for the young scientist

Robert W. Wood

Illustrated by
Steve Hoeft

Chelsea House Publishers
Philadelphia

© 1995 by Robert W. Wood.
© 1999 by Chelsea House Publishers.

Printed in the United States of America. All rights reserved. The publisher takes no responsibility for the use of any of the materials or methods described in this book, nor for the products thereof.

This Chelsea House Edition with Permission of the McGraw-Hill Companies.

Product or brand names used in this book may be trade names or trademarks. Where we believe that there may be proprietary claims to such trade names or trademarks, the name has been used with an initial capital or it has been capitalized in the style used by the name claimant. Regardless of the capitalization used, all such names have been used in an editorial manner without any intent to convey endorsement of or other affiliation with the name claimant. Neither the author nor the publisher intends to express any judgment as to the validity or legal status of any such proprietary claims.

Library of Congress Cataloging-in-Publication Data

Wood, Robert W., 1933-
    Where? : experiments for the young scientist / Robert W. Wood:
illustrated by Steve Hoeft.
      p. cm.
    Originally published : New York : TAB Books, 1995.
    Includes index.
    ISBN 0-7910-4848-9 (hardcover)
    1. Science—Experiments—Juvenile literature.
 I. Hoeft, Steve. II. Title.
Q164.W685   1997
507.8—dc21   97-27051
                                                        CIP
                                                         AC

Contents

Introduction   vii

Symbols used in this book   ix

THE YOUNG ENGINEER   1

1. Where was the first airline started?   2
2. Where was the first parachute jump made?   4
3. Where was the first instrument flight made?   8
4. Where were machines first used?   12
5. Where were drills first used?   16
6. Where can you see electricity producing light/heat?   20
7. Where can you see chemicals producing electricity?   23

THE YOUNG ASTRONOMER   25

8. Where can you find the North Star and the Big and Little Dippers?   26
9. Where can you photograph star tracks?   29
10. Where does each new day begin?   32
11. Where is the prime meridian?   35
12. Where can we get energy in space?   39

THE YOUNG CHEMIST   43

13. Where can you see soap eat an egg?   44
14. Where can you see a lemon shine a pot?   47
15. Where can you see rust form in minutes?   49

16 Where does salt come from?   53

17 Where can you see molecules mixing together (diffusion)?   55

18 Where can you see molecules separate (chromatography)?   57

THE YOUNG METEOROLOGIST   59

19 Where can you see the weight of the atmosphere?   61

20 Where can you see paper measure humidity?   64

21 Where did Celsius/Fahrenheit come from?   68

22 Where is the hottest/coldest place on Earth?   71

23 Where is the wettest/driest place on Earth?   75

24 Where is the windiest place on Earth?   79

THE YOUNG BIOLOGIST   83

25 Where do plants come from?   84

26 Where can you see how a plant eats?   87

27 Where can you see a banana ripen a tomato?   89

28 Where can you see a tadpole turn into a frog?   91

29 Where do frogs go in winter?   94

30 Where are our sense organs?   96

THE YOUNG PHYSICIST   101

31 Where can you see air pressure make heat?   102

32 Where can you see a candle burn steel?   104

33 Where can you see a needle floating in air?   106

34 Where can you see a hammer defy gravity?   108

35 Where can you see a diving medicine dropper?   110

36 Where can you see a Ping Pong ball floating in the air?   113

37   Where can you see air lifting water?   115

38   Where can you see sound patterns?   118

Appendix: Museums for seeing where science happens   121

Index   131

About the author   134

Introduction

Science covers an immense field of our knowledge. It has made our lives easier and more pleasant. Science has revolutionized the raising of crops and livestock. It has given us longer and healthier lives, and has drastically changed the way people travel and communicate with each other. Where science happens is everywhere. Many tools and techniques were invented in Egypt as early as 3000 B.C. During the second century B.C., a center of Greek learning developed in Alexandra, Egypt. Medicine was practiced to care for injured gladiators in Rome in the first century A.D. The Chinese made advances in astronomy, chemistry, and medicine, but had little contact with the outside world. The Arabs learned the decimal number system from the people of India. The Aztec, Inca, and Maya Indians of America developed a number system and had a highly accurate calendar. Science happens all around us, all the time.

This book is one of a series of books that ask such stimulating questions as "Why," "What," "When," "Who," and "Where" science happens. It provides experiments that relate to where advances have been made or where you can see science happening. At the end of the book you will find a list of almost 100 museums where you can see examples of scientific developments. Each part of the book consists of easy experiments or activities that begin by asking a question, followed by a materials list, a step-by-step procedure, and the results. At the end of each experiment you will find suggestions for further studies to broaden the scope of the experiment, followed by science trivia and oddities for your amusement.

Symbols used in this book

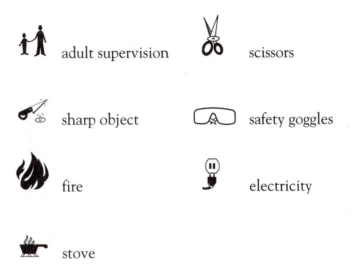

Part 1
The young engineer

Engineering is the science that applies scientific knowledge to practical uses in such fields as aeronautical, mechanical, and electrical engineering. Engineers are builders. They are often pioneers who develop advances for the betterment of people. Aeronautical engineers design the shape and structure of aircraft to extend their speed and range. Mechanical engineers work with the design of machines, the strength of materials, and how to use gears and levers to accomplish work. Advancements in the family car are good examples of engineering. Electrical and electronic engineers are responsible for most of the comforts we enjoy in our homes today.

1
Where was the first airline started?

materials ☆
- ❑ Map of Europe
- ❑ Map of the United States
- ❑ Encyclopedia

procedure ☆
1. Locate Germany on the map, then find the major cities. The world's first airline company was established as the German Airship Transport Company (Deutsche Luftschiffahrts Aktien Gesellshaft, or DELAG for short) by Count Ferdinand von Zeppelin in 1909.
2. Locate Florida on the map and then find Tampa and St. Petersburg. The world's first scheduled airline service that used airplanes was formed by Thomas Benoist and Percival Fansler in 1913.

results ☆ After surviving several near disasters, DELAG managed to establish expensive but luxurious passenger service between major cities throughout Germany. Regular schedules were never flown, and the routes depended upon who the captain was and how hard the wind was blowing. However, in the years before World War I, they did manage to provide several thousand passengers with deluxe service and an almost spotless safety record.

The first flight by Benoist Airline began on January 1, 1914. The first airline pilot was Anthony H. Jannus. The first airline passenger, A.C. Pheil, was once mayor of St. Petersburg. The airplane was a Benoist open cockpit biplane flying boat. The flight originated at St. Petersburg at 10 A.M. and landed at Tampa 30 minutes later. By 11:30 A.M., the flight had returned to St. Petersburg. Two round trips were made each day, carrying two passengers on each trip. The world's first scheduled airline service had arrived.

The first scheduled airline using airplanes made its first flight from St. Petersburg, Florida.

further studies ☆ Look up "airship" in an encyclopedia and read about the different types of airships and how they were used. Could airships become popular today? Other than advertising, how could they be used? Find "airlines" in your phone directory and notice how many companies there are and how many cities they serve.

did you know? ☆
- ❐ That passengers paid 200 marks for a two-hour flight on a German airship—a considerable price even in today's money.
- ❐ That the St. Petersburg-Tampa operation lasted only four months.

Where was the first airline started?

2
Where was the first parachute jump made?

materials ☆
- ☐ Styrofoam meat dish, about 8 × 10 inches (20.3 cm × 25.4 cm)
- ☐ Scissors
- ☐ String
- ☐ Transparent tape
- ☐ Large paper clips

procedure ☆

1. You will be making a sky diving type parachute. Make sure the Styrofoam dish has been thoroughly washed and dried. Remembering to always cut away from yourself, cut about a 6-inch (15.2 cm) arc from the long edge of the dish on one side. Next, cut about a 3½-inch (8.9 cm) arc from the opposite edge. The smaller cut will be the front of the parachute. The bottom of the dish will be the canopy. Cut the string into four equal lengths about 18 inches (45.7 cm) long.

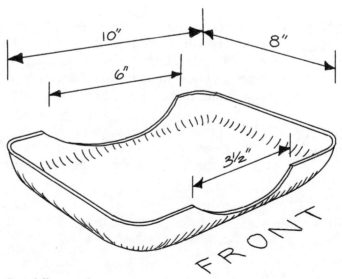

Carefully cut the curves in the front and back sides of the Styrofoam dish.

4   The young engineer

2. Tape the ends of the four strings to the four corners of the dish. Position the ends at the same location on each corner. Balance is very critical.

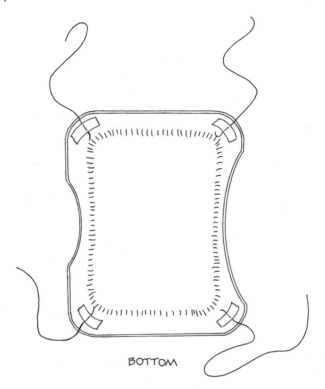

Tape the strings to the corners of the dish.

3. Holding all four strings, turn the canopy upside down and place a small weight in the center of the dish. Pull the free ends of the strings together to a point above the canopy. Viewing from the side, adjust the strings so that the two front strings are slightly shorter. This means that as you pull the strings, the point will be just a little off center toward the front of the canopy. View from the front to position the point exactly over the center of the dish. Fasten the strings together securely with tape. Attach a large paper clip to the taped point.
4. Turn the parachute right side up, hold it high with the canopy level, then let it drop.

Where was the first parachute jump made?

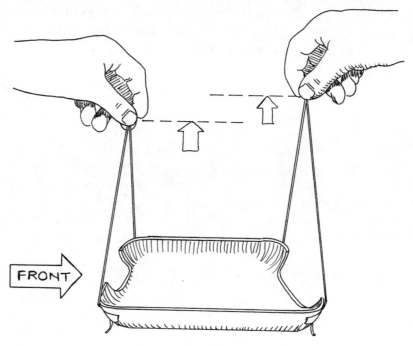

The front two strings should be a little shorter.

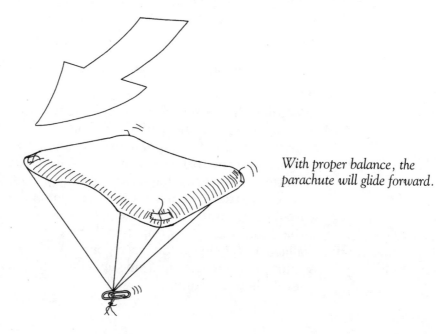

With proper balance, the parachute will glide forward.

The young engineer

results ☆ Your parachute should drop smoothly, with some forward movement. A sky diver controls the jump by shifting his weight, so anything out of balance will affect your parachute. You might have to experiment some.

Conventional parachutes have round canopies, but those used for skydiving are shaped more like an airfoil and are very maneuverable. The first live jumps were made by animals. They were released from a balloon over Vienna by Jean Pierre Blanchard in August, 1791. Blanchard made a jump in 1793 but broke his leg. The first successful jump from a balloon was made from 2,230 feet (680 m) by Andre Jacques Garnerin near Paris on October 22, 1797. The first parachute jump from an airplane in the U.S. was made from 1,500 feet (460 m) by Captain Albert Berry on March 1, 1912. The airplane was a Benoist piloted by Anthony Jannus over Jefferson Barracks, St. Louis, Missouri. The first parachute jump from an airplane by a woman was made by 18-year-old Georgia Broadwick from an airplane flown by Glenn Martin at about 1,000 feet (305 m) over Griffith Field, Los Angeles, on June 21, 1913.

further studies ☆ Find out if there are any sky-diving clubs in your area and ask about their scheduled jumps. Try to attend any events. Look up "parachute" in an encyclopedia and compare conventional parachutes to the ones used by sky divers.

did you know? ☆
- ❏ That U.S. Marine pilot William H. Rankin ejected from a Navy fighter at 47,000 feet (14,326 m) over North Carolina on July 26, 1959. Falling through a violent thunderstorm that repeatedly carried him back up, the descent took 40 minutes instead of the normal 11 minutes.
- ❏ That U.S. Air Force Captain Joseph Kittinger stepped from a balloon gondola at 102,200 feet (31,150 m) over Tularosa, New Mexico on August 16, 1960. His free fall of 84,700 feet (25,816 m) lasted a little over four and a half minutes. During the free fall, he reached a speed of 614 mph (988 kmph). His parachute opened at 17,500 feet (5,334 m). The total time for the jump was nearly 14 minutes.

Where was the first parachute jump made?

3
Where was the first instrument flight made?

materials ☆
- ❏ A couple of friends
- ❏ Blindfold
- ❏ Revolving stool
- ❏ An uncluttered room

procedure ☆
1. Study the locations of a few items in the room. Ask your friends to blindfold you and lead you safely around the room a couple of times. Now point to one of the items and remove your blindfold.

The blindfold should fit loosely over your eyes.

2. Spread your arms out and try to stand on one foot. Use your arms and your other leg to help keep your balance. Look straight ahead and focus on some nearby point. Now try it with your eyes closed.

3. Sit on the stool and locate a reference point in the room other than a window or other light source. Close your eyes and slowly turn yourself several revolutions in each direction. Point to the reference point and open your eyes.

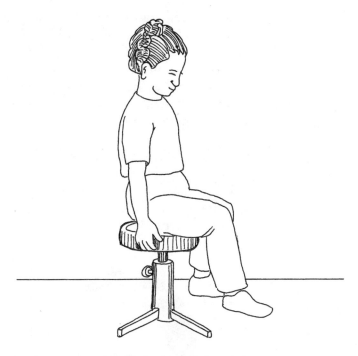

With your eyes closed, slowly turn the stool.

4. Keep your eyes open and spin yourself in one direction for several seconds then stop abruptly.

results ☆ When you were blindfolded and led around the room, was it difficult for you to keep your sense of direction? With a little practice, you should be able to stand on one foot. But with your eyes closed, wasn't it difficult to do even for a few seconds? You probably lost your sense of direction when you turned slowly on the stool, and even became dizzy when you spun faster and stopped quickly.

Our eyes are very sensitive to movement and very important for determining direction and maintaining balance. We also depend on three fluid-filled semicircular canals in our inner ears. When you tilt your head, the fluid moves much like water moves when you tilt a glass. The movement causes changes in pressure in the canals, and signals are sent to the brain. The brain reads the signals as changes in the position of the body in relation to the ground. When you are dizzy, the fluid in the inner ear continues to spin. This spinning signals the brain that you are still spinning, but your eyes tell your brain you are stopped. Imagine how hard it would be to ride a bicycle blindfolded. An airplane in flight produces gravitational forces different from those experienced on the ground. Because of these forces, it is impossible for a pilot to maintain control without the instruments that indicate reference points on the Earth. The locations of reference points are identified by radios and radar.

The first flight made without seeing the ground was made by Lieutenant James Doolittle on September 24, 1929, at Mitchell Field, Long Island, New York. He flew a biplane with an enclosed cockpit that prevented him from seeing anything except his lighted instrument panel.

James Doolittle made the first instrument flight.

further studies ☆ The next time you are a passenger in a car, close your eyes and try to keep track of your location. To investigate more about flight instruments, learn how the weather affects flight schedules at your local airport.

did you know? ☆
- ❐ That Lieutenant Ben Kelsey flew with Doolittle as a backup in case of an emergency.
- ❐ That Captain Albert Hegenberger made a blind takeoff and landing alone in a hooded cockpit on May 7, 1932, at Patterson Field, Dayton, Ohio.

Where was the first instrument flight made?

4
Where were machines first used?

materials ☆
- ☐ Pair of pliers or pry bar
- ☐ Wheelbarrow or nutcracker
- ☐ Tweezers or fishing rod

procedure ☆

1. Working with adult supervision, open and close the pliers to grip something. Notice how pressure on the handles applies a stronger force on the jaws. Use the pry bar to lift a heavy object. Find the pivot points of each tool.

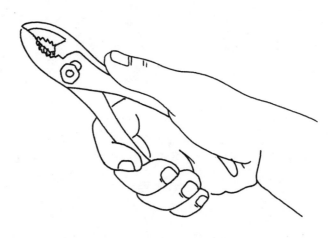

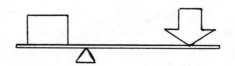

The pressure on the jaws is greater than the pressure applied to the handles.

12 The young engineer

2. Lift the handles of a wheelbarrow. Notice that the pivot point is at the wheel. Grip something with the nutcracker.

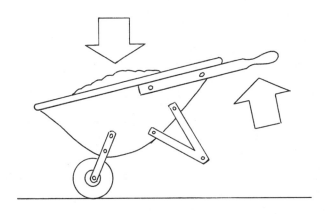

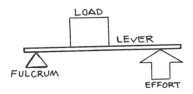

The load is balanced a little behind the wheel.

3. Open and close the tweezers, then lift an imaginary fish with the fishing rod.

results ☆ It is generally thought that humans first evolved in Africa, but some scientists believe that humans evolved in different parts of the world at about the same time. In either case, as early humans found work to do, they began to develop simple machines. One of the simplest was the lever. Early levers were used in Egypt to raise heavy buckets of water from wells. Around 200 B.C., the Greek mathematician Archimedes discovered the scientific laws of the lever.

There are three basic types of levers. The first is where the pivot point, or *fulcrum*, is between the effort and the load, as demonstrated by the pliers and pry bar. The second type of lever is

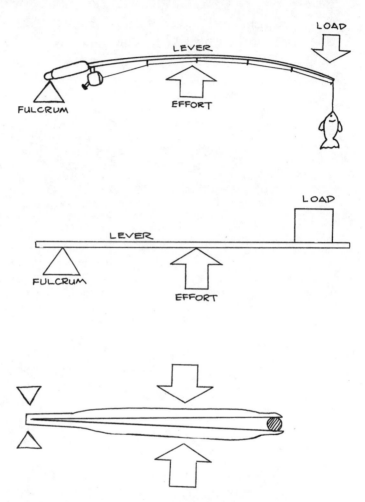

This lever is used to move the load a greater distance.

where the load is between the fulcrum and the effort, as with the wheelbarrow. The third type is where the effort to do work is applied between the load and the fulcrum, as seen in the tweezers and the fishing rod. The first two types of levers, the pliers and the wheelbarrow, are force-multiplying levers which have mechanical advantages. The third type has no mechanical advantage and is a motion-multiplying lever. In all levers, what is gained in work is lost in motion. This means that heavy loads can only be lifted a little at a time.

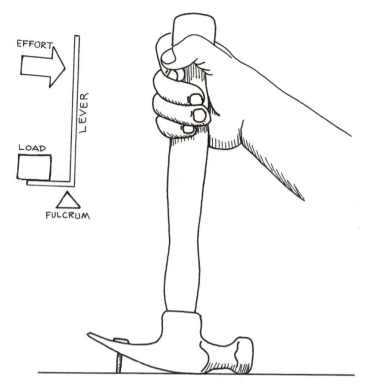

The force applied to the handle is multiplied at the claws of the hammer.

further studies ☆ Is a clawhammer a type of lever? When you pick up something with one hand, is your elbow the pivot point? Is a screwdriver or a broom a lever? Other simple machines include the wedge and the inclined plane. Are axes and loading ramps simple machines?

did you know? ☆
- ❐ That the principles of levers are also found in pulleys and the wheel and axle of a car.
- ❐ That Archimedes is supposed to have boasted "Give me a place to stand on, and I will move the Earth."

Where were machines first used?

5
Where were drills first used?

materials ☆
- ☐ Curved stick about 18 inches (25.4 cm) long
- ☐ Leather shoelace or strong twine about 2 feet (122 cm) long
- ☐ Old wooden broom or mop
- ☐ Handsaw
- ☐ Large thumbtack
- ☐ Hammer
- ☐ Nail
- ☐ File
- ☐ Shoelace
- ☐ Curved stick
- ☐ Soup spoon
- ☐ Piece of wood to drill
- ☐ Small rag

procedure ☆

1. Have an adult saw about 12 inches (30.5 cm) off the end of the broom handle. Carefully press a thumbtack into the center of the rounded end of the handle. Have an adult drive the nail part way, and straight into the cut end of the handle. Use the file to cut off the nail head.
2. Tie the ends of the leather shoelace securely to each end of the curved stick. The shoelace should be slack, not tight. Make a loop in the lace and slip it over the broom handle.
3. Fit the bowl of the spoon down over the top of the broom handle so that it is resting on the thumbtack and hold it in place. Place the point of the nail on the spot to drill and slowly push the bow back and forth. If your drill does not turn freely, adjust the length of the shoelace. If the spoon becomes hot from friction, hold it with the rag.

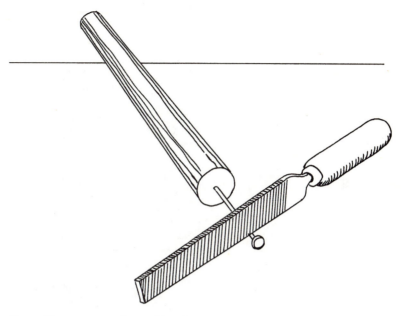

Use a file to remove the nail head.

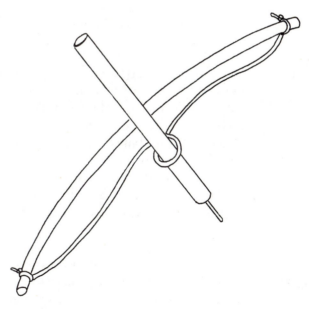

Loop the shoelace over the broom handle.

Where were drills first used?

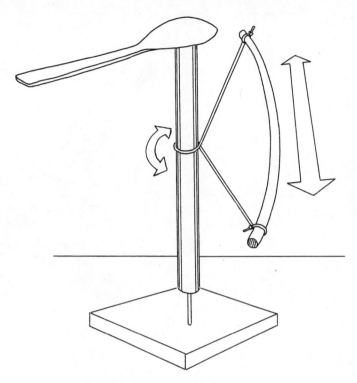

Push the bow side-to-side to turn the drill.

results ☆ With a little practice, you will be able to drill holes in most types of wood. As primitive humans developed throughout Africa and the rest of the world, they found that they wanted holes in wood and bone. Instead of nails, they used fine splinters of stone and flint to twist and make holes. They soon discovered that twisting a stick between the palms of their hands could produce fire. Later, they fastened the flint to a stick and twisted the stick between the palms of their hands to drill holes. This drilling method was improved by using a bow and bowstring much like the one you constructed.

further studies ☆ How many ways are drills used today? Are drills used in soil studies, or in highway construction? Consider the engineering feat of building the tunnel under the English Channel. The Channel Tunnel, or "Chunnel," was a joint project connecting three tunnels between England and France.

The young engineer

did you know? ☆

- That the dental drill was invented by John Greenwood in New York City in 1790. Power to the drill was supplied by a foot pedal.
- That when building the tunnel under the English Channel, English and French crews drilling from each country broke through to meet on December 1, 1990.
- That the worlds deepest undersea tunnel is the Seikan Tunnel, a railroad tunnel between the islands of Hokkaido and Honshu, in Japan.

6
Where can you see electricity producing light/heat?

materials ☆
- ❏ Old lamp cord with plug about 18 inches (46 cm) long
- ❏ Knife or wire strippers
- ❏ Thin, steel wire about 4 inches (10.2 cm) long
- ❏ Clear drinking glass
- ❏ Dark room
- ❏ Five D-cell flashlight batteries
- ❏ Cardboard tube from paper towels

procedure ☆

1. Separate the two halves of the lamp cord and carefully strip about ½ inch (1.3 cm) of the insulation from each end. Ask an adult for help if you are unsure about stripping the wire.

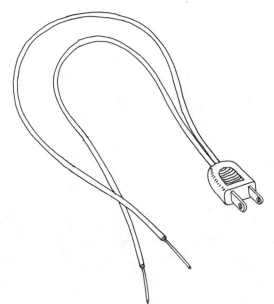

Remove the insulation from the end of each wire.

2. Tightly twist the ends of the steel wire to each prong of the plug to form a loop between the prongs. Lower the plug about halfway into the glass. Bend the cord at the rim of the glass to hold the plug in position.

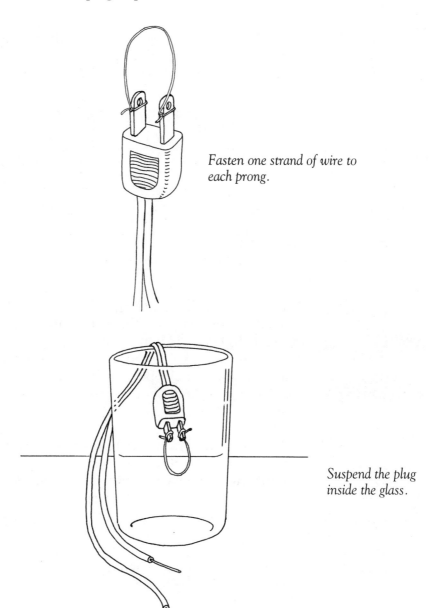

Fasten one strand of wire to each prong.

Suspend the plug inside the glass.

Where can you see electricity producing light/heat?

3. Ask an adult to slide the batteries into the tube all in the same direction. Turn out the room lights and ask an adult to touch the bare ends of the lamp cord to each end of the tube of batteries.

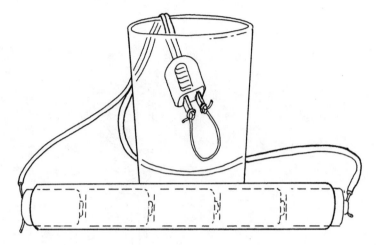

Touch the bare ends of the wires to the batteries.

results ☆ The wire will get hot and begin to glow producing heat and light. Soon, the wire will get so hot that it will probably break. In a regular light bulb, the bulb is filled with nitrogen and the wire filament is made of tungsten. This allows the bulb to last longer. You can find electricity producing light and heat all around the world. Can you think of other examples of electricity producing light?

further studies ☆ Hold your hand a few inches below a table lamp and turn on the lamp. You should feel the heat almost as soon as the bulb lights. Does lightning produce heat and light from electrical energy? Why does a fluorescent light give off less heat than a regular incandescent light bulb? Do incandescent lights waste energy?

did you know? ☆
❏ That at any time around the world, about 2,000 thunderstorms may be producing 100 lightning flashes each second.
❏ That lightning bolts begin with at least 10,000 volts of electricity and can have a current of 20,000 amps, enough energy to power a small town for a year.

7
Where can you see chemicals producing electricity?

materials ☆
- ☐ Stiff steel wire, nail or paper clip about 4 inches (10.2 cm) long
- ☐ A large potato
- ☐ A lemon, or lemon juice
- ☐ Stiff, copper wire about 4 inches (10.2 cm) long
- ☐ Direct current (dc) voltmeter with connecting leads

procedure ☆

1. Insert one end of the steel wire into one end of the potato. Stick the copper wire into the other end of the potato. A lemon or a small glass of lemon juice can be substituted for the potato.
2. Have an adult connect the negative lead from the meter to the steel wire, and the positive lead to the copper wire.

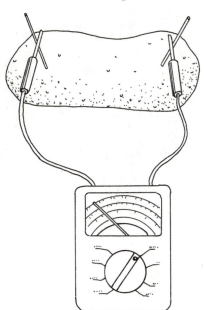

Touch the negative lead to the steel wire, the other lead to the copper wire.

results ☆ The meter should read about ½ volt. A small, steady (d.c.) electric current can be produced when two unlike metals are immersed in an acid solution. The two unlike metals in your experiment were the steel and copper wires. The juice inside the potato or lemon supplied the chemical action to produce the electricity.

further studies ☆ Where else can you find chemicals producing electricity? Do the batteries in a flashlight or the battery in a car use chemicals? What is an electrolyte? What is (d.c.) direct current?

did you know? ☆ ❒ That in 1780, in Italy, when Luigi Galvani produced an electrical spark between a dead frog's legs and a metal stand, he thought he had discovered an electrical source in animals. Alessandro Volta thought electricity came from the difference in two metals. Both were right: the chemicals in the frog's legs and in the two different metals produced the electricity.
❒ That a magnetic field is always found around an electrical current.

Luigi Galvani

Part 2
The young astronomer

Astronomy, probably the oldest of the sciences, began with humans wondering about the Sun, Moon, stars, and other objects they saw in the night sky. Astronomy began to develop between 1800 B.C. to 400 B.C. when the Babylonians invented a calendar based on the movements of the Sun and the phases of the Moon. Later, in the middle 1500s, Nicolaus Copernicus explained that the Sun, not the Earth, was the center of our universe. Early astronomers had to observe the heavens with their naked eyes until 1609, when Galileo constructed the first telescope. A reflecting telescope with a mirror 200 inches (5.08 m) in diameter was completed in 1948 at Palomar Observatory in California. In April 1990, the Hubble Space Telescope was launched into space to see ten times deeper into the universe than ever before. Problems did occur at first; however, astronauts were able to make repairs in December 1993. Advances have been so rapid in recent years that we can only guess what new discoveries will be made.

8
Where can you find the North Star and the Big and Little Dippers?

materials
- ❒ Magnetic compass
- ❒ Clear night

procedure
1. Face the northern half of the sky and look for patterns in the brightest stars. Imagine drawing lines between some of these stars.
2. Find seven bright stars that look like the side view of a pan with a handle. In the winter, the handle will be pointing down. In the summer, the handle will be pointing up.
3. Look at the two bright stars that make up the front of the dipper. Notice the space between these two stars. Measure about five of these spaces on a line to locate another star. This star will be one of the stars in a group of stars that make up a smaller dipper.

results
The first group of stars made up the constellation Ursa Major, or the Big Dipper. The two stars at the end of the Big Dipper are the pointer stars (Merak and Dubhe) that point to *Polaris*, or the North Star. The North Star is the last star in the handle of the Little Dipper. In the spring, the Big Dipper is upside down and seems to be pouring into the Little Dipper. In the autumn, the Little Dipper appears to be pouring into the Big Dipper.

further studies
Look up "astronomy" in an encyclopedia and find the constellations in the polar group. See if you can identify Cassiopeia and Draco. On a winter night, locate the constellation Orion. Next to the Big Dipper, Orion is the best-known constellation around the world. Can you see a difference between the colors of the stars Betelgeuse and Rigel?

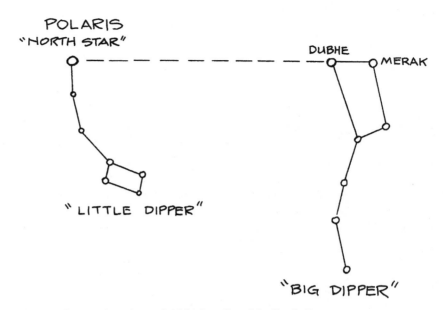

The North Star is at the end of the handle of the Little Dipper.

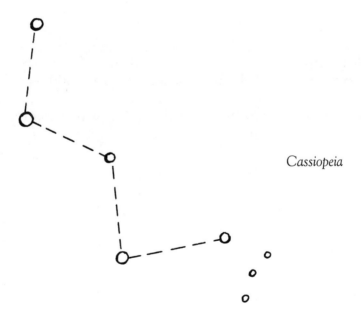

Cassiopeia

Where can you find the North Star and the Big and Little Dippers?

O BETELGEUSE

Orion

O RIGEL

did you ☆ know?

- ❐ That the light you see from Betelgeuse is 300 years old, and the light from Rigel is more than 500 years old.
- ❐ That because Earth's axis slowly changes every 26,000 years, the North Star will only be the pole star for a few hundred more years.

9
Where can you photograph star tracks?

materials ☆
- ❏ Camera with an adjustable shutter speed
- ❏ Tripod or solid platform
- ❏ Black-and-white film
- ❏ Clear night

procedure ☆
1. Load the film into the camera and find a location away from streetlights or other lights that will interfere with your view. Try to pick a night with no Moon.
2. Mount the camera on the tripod or any solid platform. Set the shutter speed so that the shutter will stay open for 30 minutes to one hour. It is very important to keep the camera steady during the exposure.

Be sure to hold the camera steady while taking the photograph.

3. For an interesting pattern of star tracks, face north and aim the camera at the North Star. Center the North Star in the view finder, take the picture, then have the film developed.

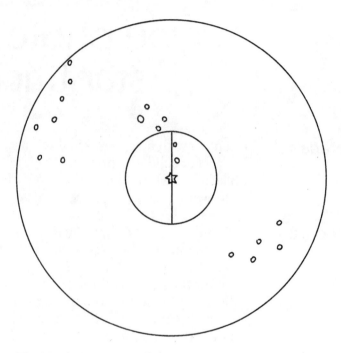

The North Star centered in the view finder.

results ☆ You can photograph star tracks any place in the world on a clear night free of surrounding lights. Your film should produce a circular pattern of star tracks around the North Star. The pattern will show that the stars seem to move in a counterclockwise direction around the North Star. You will notice that some of the constellations in the northern sky never set.

further studies ☆ Aim your camera in other parts of the sky and compare the different star tracks. Do the tracks run from east to west? Aim your camera due east and take a photograph of the tracks. Measure in degrees the angle of the tracks and, knowing that the equator is at zero degrees, compare it to the latitude (marked on a world map) where you live.

The young astronomer

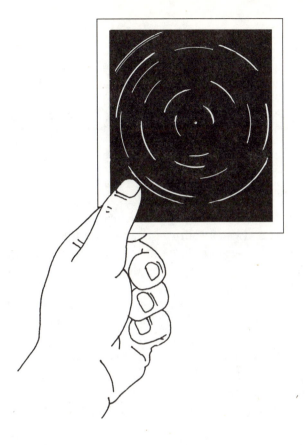

A circular pattern of star tracks.

did you ☆ know?

☐ That if you were at the equator, the star tracks would go straight up.
☐ That you cannot see the North Star if you live below the equator.

Where can you photograph star tracks? 31

10 Where does each new day begin?

materials ☆ ☐ Globe of the Earth or world map

procedure ☆ 1. Notice the equally spaced lines on the map running north and south. These are lines of longitude measured in degrees east and west. Look at the area in the middle of the Pacific Ocean and find the line marked 180 degrees.

results ☆ The 180-degree line, or meridian, is called the *international date line*. It might be marked Monday on the west side of the line and Sunday on the east side. This imaginary line was established by an international agreement in 1884 as the place where each new day begins. Our Earth day starts on the west side of the line and ends on the east side. As the Earth rotates, the new day sweeps westward, covering the entire Earth in 24 hours.

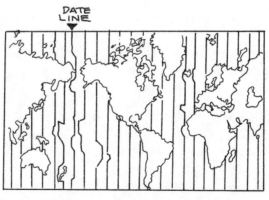

The international date line.

The young astronomer

further studies ☆ If it takes 24 hours for the Earth to complete one revolution on its axis, how many degrees does the Sun seem to travel in one hour? What is the difference in time between London and New York City? How many time zones are in the United States?

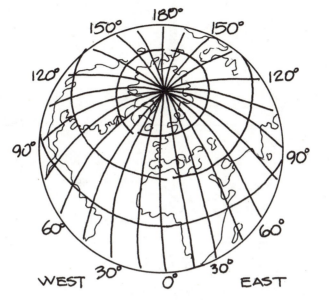

Time zones are marked off in degrees.

did you know? ☆
- ❏ That if you lived in New Zealand, you could celebrate New Year's Day 22 hours ahead of the people in Hawaii.
- ❏ That where the 180-degree meridian crosses land, the international date line is moved to one side or the other to keep from disrupting the pattern of life of the people in the area.

Where does each new day begin?

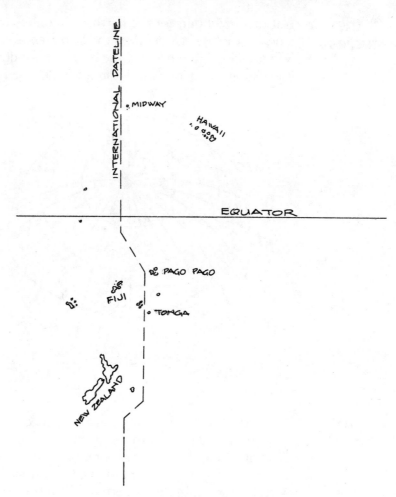

New Zealand and Hawaii are on opposite sides of the date line.

11
Where is the prime meridian?

materials ☆
- ❏ Two straight sticks about 12 inches (30.5 cm) long
- ❏ Hammer
- ❏ Globe of the Earth or world map
- ❏ Sunny day

procedure ☆

1. Have an adult drive one stick into the ground at an angle that points directly at the sun. The stick should not make a shadow. Wait about an hour, or until the stick has a shadow about 6 inches (15.2 cm) long. The shadow should be pointing east from the stick.

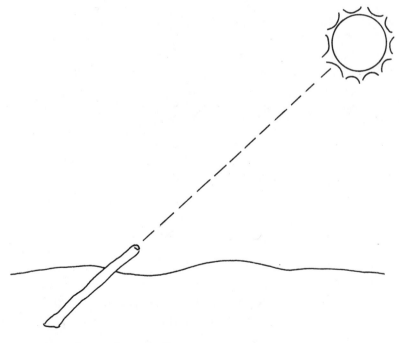

Point the stick straight at the Sun.

2. Stand with your right shoulder pointing with the shadow (east), and you will be facing north.

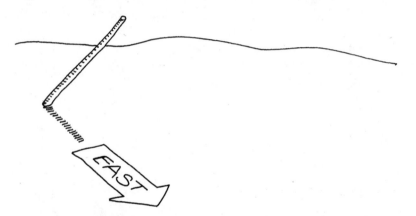

The shadow will point east.

3. Now that you know which is east and west, draw a line in the dirt running north and south. Remove the stick from the ground. Just before noon, drive both sticks straight into the ground on this line. Watch the shadows of both sticks until they are in line. This is your meridian.

4. Find London, England on the map and locate the 0-degree longitude line. Notice the River Thames running east and west.

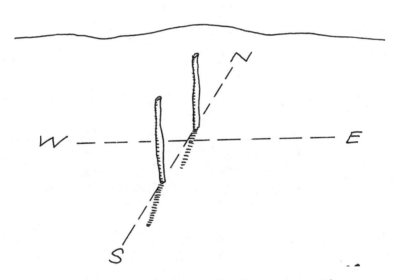

When the shadows are in line, they will indicate your meridian.

results ☆ You located the north-south line by the Sun, then found your meridian, or *longitude*, when the Sun passed precisely overhead.

On the map you found the 0-degree longitude where it passes through London. Greenwich is located on the south bank of the

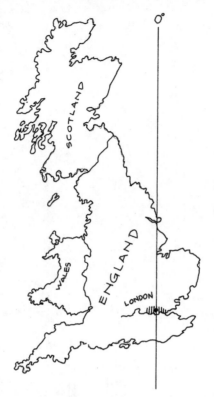

The prime meridian is 0-degrees longitude.

River Thames and was the site of the Royal Greenwich Observatory from 1675 to 1958. It is now part of the National Maritime Museum. The prime meridian is the longitude line of 0 degrees where it passes through Greenwich Observatory. It was agreed by international treaty in 1884 that time zones would be based on this line.

further studies ☆ Compare the prime meridian with the international date line. Are they on opposite sides of the Earth? How many hours does it take for the Sun to pass from London to the international date line? How many degrees is this at 15 degrees per hour?

did you know? ☆
- ❏ That the old Greenwich Observatory building has a brass strip marking the prime meridian.
- ❏ That near the equator, the speed of the Earth spinning on its axis is close to 1,000 miles per hour. The Earth makes one complete revolution every 24 hours and makes a little more than 365 revolutions in the time it takes to travel just once around the Sun.

12
Where can we get energy while in space?

materials ☆
- ❏ Candle
- ❏ Matches
- ❏ Pliers
- ❏ Two shiny metal cans the same size
- ❏ Two thermometers
- ❏ Water (cold and warm)

procedure ☆

1. Have an adult light the candle and use the pliers to grip the can. Have an adult hold the can in the flame and blacken the outside of one of the cans. Let the can stand untouched until cool.

Blacken one of the cans with soot.

Where can we get energy while in space?

2. Place a thermometer in each can and fill both cans about two-thirds full of cold tap water. Set the cans side by side in warm sunlight and monitor the changes in the temperature.

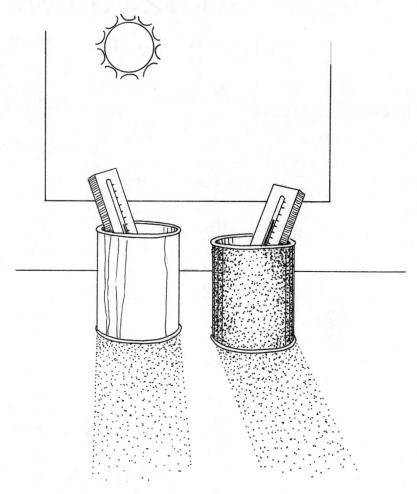

Place both cans in warm sunlight.

results ☆ In space, sunlight can be easily converted into heat energy. The water in the dark can will begin to warm first. The radiant heat from the Sun strikes both cans equally, but the dark can absorbs most of the heat while the shiny one reflects the heat.

further studies ☆ Empty each can and fill them both with warm tap water. Place them side by side in the shade and watch the changes in the temperature. Does the water in the dark can start to cool first? Does heat radiate faster from the dark surface? Can radiant energy travel through a vacuum? Can this energy generate electricity?

did you know? ☆
- ❐ That a surface coated with carbon black or soot will absorb 97 percent of the energy striking it.
- ❐ That light and radio waves are electromagnetic waves that contain radiant energy.

Where can we get energy while in space?

Part 3
The young chemist

Chemistry is the science that studies the makeup of things and the changes that occur within these substances. Chemical change is all around us. The food you eat is chemically changed inside your body to give you energy to grow and to think. Early humans used chemistry when they built fires to cook food, used heat to make bricks, and later to work with metals such as iron, copper, and gold. By the seventeenth century, chemistry was considered a science. At the beginning of the nineteenth century, John Dalton developed the idea that elements were made up of tiny particles called atoms. Today, with the aid of modern technology, new discoveries are constantly being found, and because chemistry covers the entire material universe, it is important for understanding the other sciences.

13
Where can you see soap eat an egg?

materials ☆
- ❏ Two jars
- ❏ Hard-boiled egg
- ❏ Powdered laundry detergent with enzymes
- ❏ Powdered laundry detergent without enzymes (found in detergents used for baby laundry)
- ❏ Warm tap water
- ❏ Pen and masking tape for labels
- ❏ Tablespoon and knife

procedure ☆

1. Ask an adult to help you boil an egg. While the egg is boiling, put a tablespoon of detergent containing enzymes into one of the jars and label it "enzyme." Put an equal amount in the other jar and label it "no enzyme."

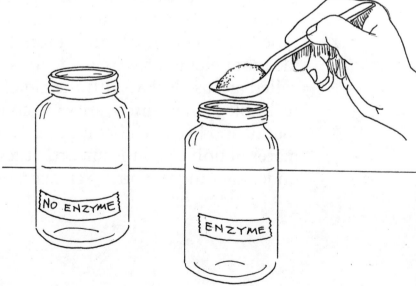

Pour a tablespoon of each type of detergent into each jar.

2. Put eight tablespoons of warm water in each jar.
3. After the egg is boiled and cooled, peel it, then ask an adult to carefully cut two pieces of the egg white exactly the same size. Place one piece in each jar.

Cut each piece of egg white the same size.

4. Place both jars in a warm location, such as near a heater vent. Let the jars sit for a couple of days, then compare the sizes of each piece.

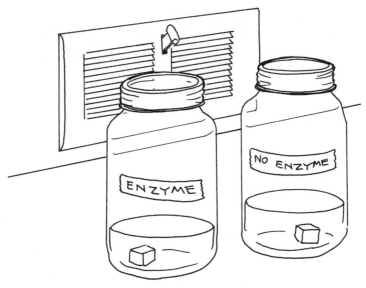

Place both jars in a warm location for a couple of days.

Where can you see soap eat an egg?

results ☆ After a couple of days, the egg white in the jar labeled "enzyme" should be smaller. Enzymes are protein-like substances formed in plant and animal cells. The enzyme attacks the particles in the egg white and breaks them into smaller particles. The smaller particles then dissolve in the water. The piece of egg in the other jar should have remained the same size because it contained no enzymes.

did you know? ☆
- ❒ That the enzymes in our stomachs break down food into molecules small enough to be carried by our blood.
- ❒ That without enzymes, chemical reactions in your cells would happen too slowly to sustain life.

14
Where can you see a lemon shine a pot?

materials ☆
- ❏ Old, dull aluminum pot
- ❏ Lemon
- ❏ Knife
- ❏ Water
- ❏ Stove

procedure ☆

1. Ask an adult to help you boil some water in the aluminum pot.
2. When the water is boiling, have an adult cut a couple of slices of lemon and drop them into the water.

Drop the lemon slices into the boiling water.

results ☆ The inside of the pot will begin to shine. Aluminum exposed to the air forms a thin coating of aluminum oxide, which protects the metal and keeps it from rusting. However, chemicals in water can cause discolorations after a period of time. When you added the lemon pieces, citric acid from the lemon mixes with the aluminum oxide coating to form *aluminum citrate*. The layer of aluminum citrate dissolved in the boiling water and exposed the shiny metal.

further studies ☆ Do oranges or grapefruits contain citric acid? Other than cooking utensils, where can you find aluminum being used in your home? Is aluminum a metallic chemical element?

did you know? ☆ ❏ That aluminum can be recycled for less than 5 percent of the energy used to produce new metal.
❏ That the United States is the largest producer and consumer of aluminum in the world.

15
Where can you see rust form in minutes?

materials ☆
- ❏ Small drinking glass
- ❏ Warm tap water
- ❏ Table salt
- ❏ Tablespoon
- ❏ Scissors
- ❏ Two copper wires (bell wire about 18 inches 46 cm) long
- ❏ Two nails
- ❏ 9-volt battery

procedure ☆

1. Fill the glass with about two inches (5.1 cm) of warm water and stir in a couple of tablespoons of salt.

Stir the salt solution.

2. Use the scissors to trim about two inches (5.1 cm) of the insulation from the ends of both wires.

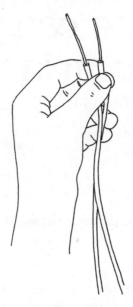

Remove the insulation from the ends of the wires.

3. Twist one of the bare ends of one wire tightly around one of the nails just below the nail head. Ask an adult to wrap the other bare end of this wire around one of the terminals of the battery.

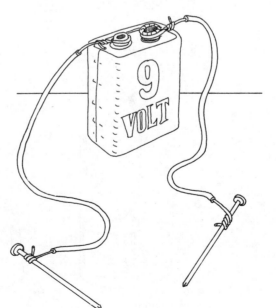

Connect the wires to the battery.

Wrap the other wire to the other nail the same way and have an adult connect the free end of the wire to the remaining terminal on the battery.

4. Keeping the nails separated, have an adult lower them both into the salty solution. Notice which nail is connected to the positive terminal of the battery. After a few minutes remove the nails and examine them.

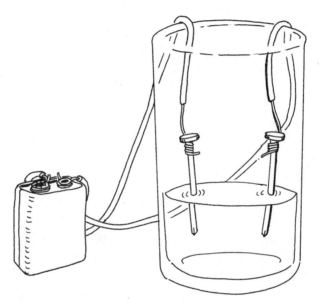

Lower the nails into the salt solution.

results ☆ When they were lowered into the solution, bubbles immediately started forming around one of the nails and not the other. The nail connected to the negative terminal of the battery will still be shiny, while the one connected to the positive terminal will have started to rust. The shiny nail with the negative polarity attracted the hydrogen in the solution and formed bubbles. The hydrogen bubbles kept the nail from rusting. The nail with the positive polarity attracted chlorine from the solution, which attacked the nail and caused rapid corrosion, or rust.

further studies ☆ Is oxygen and moisture necessary for corrosion to take place? What is oxide and how does it protect aluminum from corrosion? When moisture is present, does iron become porous to oxygen? Is corrosion a problem in dry climates?

did you ☆ ❑ That in the United States corrosion causes over $10 billion in damages every year.
know?
❑ That, because of corrosion, paint crews are constantly repainting the Golden Gate Bridge in San Francisco.

16
Where does salt come from?

materials ☆
- ☐ Tablespoon
- ☐ Sink
- ☐ Vinegar
- ☐ Baking soda

procedure ☆ 1. Hold the spoon over the sink and fill it about half full of vinegar.

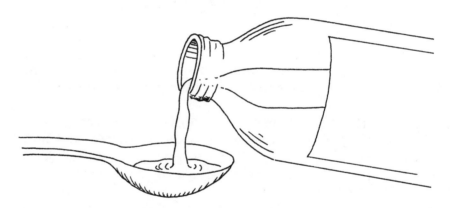

Pour a little vinegar into the spoon.

2. Sprinkle baking soda on top of the vinegar until the bubbling stops.

results ☆ The mixture has become a salt. Vinegar is an *acid* and soda is a *base*. A base is the opposite of an acid. The bubbles were formed by carbon dioxide gas. When you add a *soluble* (able to be dissolved) base to an acid, you will get a salt. A base that dissolves in water is called an *alkali*. The acid and the alkali react together to form the salt. Table salt, Epsom salts, seawater salt, and bath

Add baking soda to the vinegar.

salts are only a few of the compounds called salts. Most minerals are salts, and all water contains some dissolved salts.

further studies ☆ Except for seasoning foods, what other uses can you think of for salt? Can salt be used to preserve food? Can salt be used when making homemade ice cream or to melt ice on streets and sidewalks? Is blood a saline solution?

did you know? ☆
- That too much salt intake in your body can cause hypertension, also known as high blood pressure.
- That fields along the Nile River were successfully irrigated for 5,000 years. Natural annual flooding washed the salt buildup due to irrigation of the fields, but since the completion of the Aswan High Dam, the fields and the Nile River itself have been increasing in salt content.

17
Where can you see molecules mixing together (diffusion)?

materials ☆
- ☐ Glass of cold tap water
- ☐ Red food coloring

procedure ☆
1. Drop a couple of drops of food coloring into the water. Do not stir the water. What happens to the food coloring?
2. Let the water stand overnight. What color is the water?

Add a couple of drops of food coloring to the water.

results ☆ At first, the drops of food coloring formed something like pink clouds. But after setting overnight, the water was tinted completely because the food coloring had spread throughout the water. The molecules of food coloring and the molecules of water mixed together due to a process called *diffusion*.

Atoms are joined together by chemical bonds to make up molecules. Diffusion is a process where molecules move from a higher concentration into an area of lower concentration. Diffusion will continue to occur without stirring, until the mixture is equal throughout.

further studies ☆ Diffusion can occur in liquids. Can it take place in gases or solids? Does diffusion take place if a strong odor is released in one part of a room? Does diffusion take place faster in gases than in liquids?

did you know? ☆
- ❏ That molecules can contain from two to hundreds of thousands of atoms. A molecule of oxygen for example, has two atoms.
- ❏ That before the eighteenth century, experiments with liquids and solids were easily handled, but experiments with heated gases had to be done with animal bladders.

18
Where can you see molecules separate (chromatography)?

materials ☆
- ❐ Scissors
- ❐ Paper towel or toilet tissue
- ❐ Different colors of food coloring
- ❐ Mixing jar
- ❐ Water
- ❐ Small, clear drinking glasses or jars
- ❐ Paper clips

procedure ☆
1. Cut the paper into strips. Mix two or more types of food coloring together in the mixing jar. Place a drop of the mixture on one end of a strip of paper.

Drop a small amount of the solution on one end of the paper.

2. Rinse the mixing jar and mix two other colors together. Place a drop of the mixture on another strip of paper. Repeat the steps with other colors.
3. Pour about an inch (2.5 cm) of water into each glass. Lower the strips into each glass so that the colored drops are just below the surface of the water. Fasten each strip over the rim of the glass with a paper clip. Monitor the color changes of the drops for about 30 minutes, then remove the strips and allow them to dry. What happened to the colored drops?

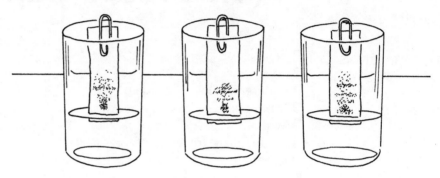

Place the ends with the colored drops in the water.

results ☆ The colored drops will migrate up the strips of paper and partially separate into different colors by a process called *paper chromatography*. The process works through *adsorption*, where the molecules of the dissolved liquid attach themselves to the surface of the paper. Because each type of molecule is adsorbed at a different rate of time, they separate as they move up the strip of paper.

further studies ☆ Repeat the experiment and record the time it takes for the water to carry the colors up the paper. Which colors move fast, and which ones move slow? Mix several colors together and try different combinations of colors. Can chromatography be used to separate and identify the components of a mixture?

did you know? ☆ ❑ That chromatography was first used to separate the pigments in plants.
❑ That chromatography comes from the Greek word meaning "color writing."

Part 4
The young meteorologist

The word "meteorology" comes from the Greek word *meteoron*, which refers to anything that happens in the sky. In ancient times, Aristotle's "meteorological" even included the study of meteors and astronomy in general. Today, meteorology is considered to be the study of the atmosphere, as well as the changes in temperature and moisture that produce various weather conditions. Meteorologists study the conditions that produce rain, snow, thunderstorms, and severe weather such as tornadoes and hurricanes.

Weather affects us all in various ways. Low moisture conditions can cause crop damage, forest fires, and water shortages in cities. High moisture conditions can produce floods and other hardships. The importance of weather on our lives led to the development of the science of weather forecasting. Meteorologists use high-speed computers to handle vast amounts of information received from satellites and ground-based stations. Important information can then be passed around the world through a network organized by the World Meteorological Organization. It is easy to see that meteorology is a science we all depend on.

19
Where can you see the weight of the atmosphere?

materials ☆
- ☐ Water
- ☐ Measuring cup
- ☐ Gallon can with screw cap
- ☐ Stove
- ☐ Protective gloves or pot holders
- ☐ Sink

procedure ☆

1. Pour one cup of tap water into the can and leave the cap off. Place the can on the stove. Ask an adult to help you heat the water to a boil.

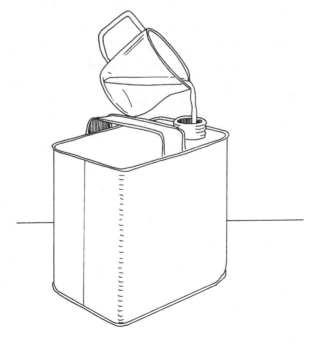

Pour a cup of water into the can. Then leave the cap off while heating the water.

2. Using gloves, carefully remove the can from the stove then quickly screw on the cap. Place the can in the sink and run cold water over it. What happens to the can?

After heating the water, install the cap and run cold water over the can.

results ☆ The can instantly collapsed. Our atmosphere surrounds the Earth and is pulled down by gravity just like solids and liquids. Atmospheric pressure is measured in "atmospheres." An atmosphere is a pressure of 14.7 pounds per square inch of surface. When the boiling water heated the air inside the can, the air expanded creating a very low pressure. With the cap in place, the cold water chilled the air inside the can creating a partial vacuum, allowing the normal atmospheric pressure of almost 15 pounds per square inch to crush the can.

further studies ☆ Measure the approximate area of the surface of the can and estimate how much pressure was applied to the can. Is this same pressure pressing on you? Hold your hand flat and guess how much pressure is pushing down on it. Why can't you feel the weight? Is the air pressure pushing on the bottom of your hand at the same time?

did you ☆ know?

- That simple lift pumps used to pump water from wells cannot lift water any more than about 34 feet (10 m). Atmospheric pressure is only great enough to push water up about 34 feet (10 m) when the air is removed from the well pipe by the pump.
- That one atmosphere measures 29.92 inches (760 mm) in a barometer of mercury.
- That around 200 B.C., the Greeks built an observatory called "The Tower of Winds" to study the atmosphere

20 Where can you see paper measure humidity?

materials ☆
- ☐ Scissors
- ☐ Aluminum foil
- ☐ Sheet of newspaper
- ☐ Transparent tape
- ☐ Pencil
- ☐ Wooden spool from sewing thread

procedure ☆
1. Cut a piece of aluminum foil about 4 inches (10. cm) wide and 10 inches (25.4 cm) long. Cut a strip of newspaper about 1 inch (2.54 cm) wide and 10 inches (25.4 cm) long.

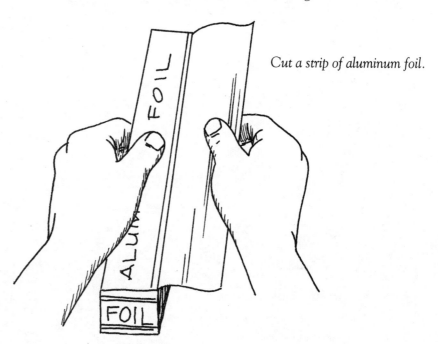

Cut a strip of aluminum foil.

2. Place the strip of newspaper down the middle of the aluminum foil and fasten just the edges of the paper in place with tape.

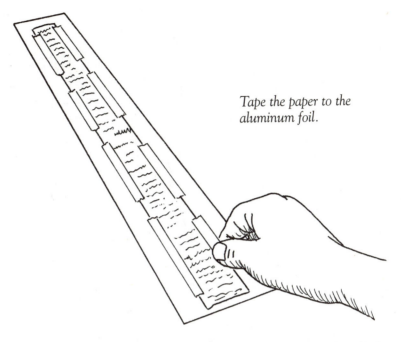

Tape the paper to the aluminum foil.

3. Trim each edge of the foil so that it is about 1½ inches (3.8 cm) wide. You should now have a strip of foil 1½ inches (3.8 cm) wide and 10 inches (25.4 cm) long with a strip of newspaper on one side.
4. Tape one end of the strip near the top of the pencil and wind it, with the paper on the inside, snugly around the pencil. Stand the pencil upright in the spool and place it in a moist place, such as a bathroom. Observe the strip before and after moisture has been in the room. What does the strip do?

results ☆ When moisture enters the room, the strip will unwind. When the air becomes dryer, the strip will wind back up. The paper absorbs moisture and expands. The aluminum foil cannot. This forces the curl to unwind and wind back up as the moisture content in the air changes.

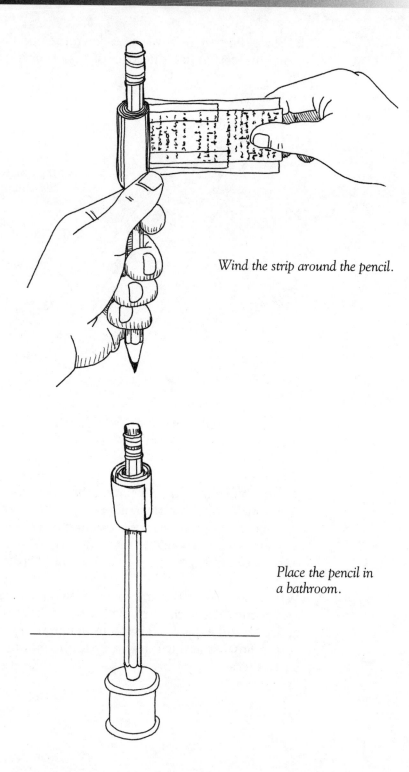

Wind the strip around the pencil.

Place the pencil in a bathroom.

The young meteorologist

further studies ☆ What other materials are affected by humidity? When the humidity is high, does it affect how your hair behaves? Does humidity affect how we feel? What is a hygrometer?

did you know? ☆
- ❏ That when a weather report gives a relative humidity of 50 percent, it means that the air contains half of the moisture it can hold.
- ❏ That in 1775, Horace de Saussure thought of using human hair to measure humidity because it stretches and shrinks when humidity increases and decreases. A hair is fastened to a pointer that moves across a scale.

21 Where did Celsius/Fahrenheit come from?

materials ☆
- ❏ Fahrenheit thermometer
- ❏ Celsius thermometer
- ❏ Two cups or glasses
- ❏ Crushed ice

procedure ☆

1. Take your temperature by placing each thermometer under one of your armpits. Compare the readings of the thermometers.

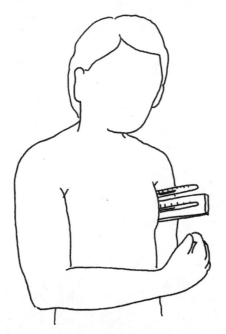

Measure the temperature of your body.

2. Place a thermometer in each cup and fill the cups with ice. Let the cups sit for three or four minutes, then read the temperature of each thermometer. How do the readings compare?

Place the thermometers in ice.

results ☆ The temperature under your arm is normally about one degree Fahrenheit lower than your actual temperature, so add 1 degree F and 0.5 degree C to the readings. You should have a body temperature of 98.6 degrees F and 37 degrees C. The temperature in the cups should be about 32 degrees F and 0 degrees C. The *Fahrenheit* scale is widely used, although not often in scientific measurement. It was introduced in 1720 by Gabriel Fahrenheit. The *Celsius* scale, sometimes called the *centigrade* scale, was adopted in 1948 in honor of Anders Celsius.

further studies ☆ Fahrenheit degrees are smaller than Celsius degrees. One Fahrenheit degree equals 5/9 of a Celsius degree. To change Fahrenheit to Celsius, subtract 32, multiply by 5, and divide by 9. To change Celsius to Fahrenheit, multiply by 9, divide by 5, and add 32. Practice converting the temperatures of your readings and see how they compare. What is the freezing point of water and the boiling point of water on each scale? Is the Celsius scale ever used in scientific measurements?

Where did Celsius/Fahrenheit come from?

did you know? ⋆

- ❐ That Gabriel Fahrenheit manufactured weather instruments in Amsterdam and was the first to introduce a practical mercury thermometer.
- ❐ That Anders Celsius was a Swedish astronomer, and that he went to Lapland to make measurements to prove Newton's theory that the Earth was slightly flatter at the poles.

22
Where is the hottest/coldest place on Earth?

materials ☆
- ❏ Lamp
- ❏ Thermometer
- ❏ World map
- ❏ Freezer compartment in refrigerator

procedure ☆ 1. Turn on the lamp and place the thermometer near the bulb. Locate the Libyan desert on the map, then check the reading on the thermometer.

Measure the temperature near a light bulb.

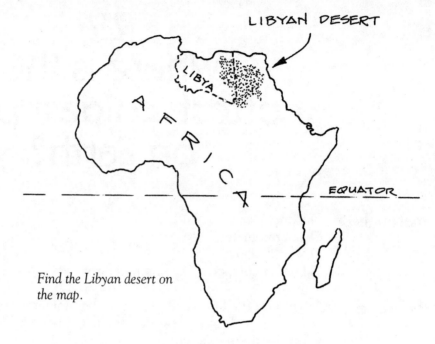

Find the Libyan desert on the map.

2. Allow the thermometer to cool some, then place it in the freezer compartment of the refrigerator. Locate Antarctica on the map, then check the reading on the thermometer.

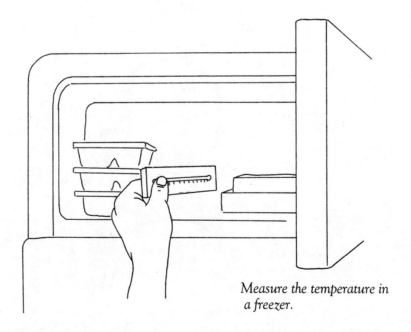

Measure the temperature in a freezer.

The young meteorologist

Find Antarctica on the map.

results ☆ A typical thermometer might have a scale from minus 40 degrees to 120 degrees F. The temperature near the lamp was probably higher than the thermometer could show. The highest temperature ever recorded was in the Libyan desert with a temperature of 136 degrees F (57.8 degrees C). The lowest was recorded in Antarctica at a minus 128.6 degrees F (minus 89.2 C). The temperature in the freezer compartment of the refrigerator might have been about 4 degrees F.

further studies ☆ Do you think people could live and work in these extreme temperatures? What types of vehicles could operate? In extreme cold, would car tires become brittle and break instead of going flat? What types of plants and animals would you expect to find?

did you know? ☆
❏ That the year-round average temperature in the shade at Dallol, Ethiopia is 94 degrees F (34.4 degrees C), the hottest average in the world.
❏ That the coldest is at Vostok, Antarctica with an average temperature of minus 72 degrees F (minus 57.8 C).

❐ That the greatest temperature change in the United States in one day occurred at Billings, Montana when the temperature dropped from 46 degrees (7.8 C) to minus 55 degrees F (minus 48.3 C), a change of 101 degrees F (56.5 C).

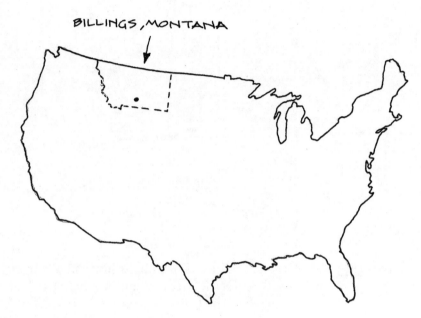

Billings, Montana, recorded the largest temperature change in the United States.

23
Where is the wettest/driest place on Earth?

materials ☆
- ❏ Can with straight sides (coffee can)
- ❏ Water
- ❏ Funnel
- ❏ Tall jar with straight sides (olive jar)
- ❏ Ruler
- ❏ Marking pen
- ❏ Watch or clock
- ❏ World map

procedure ☆ 1. Place the can on a level surface. Pour 2 inches (5.08 cm) of water into the can. Use the funnel to pour the water from the can into the tall jar. Mark the level of the water on the outside of the jar, then pour out the water.

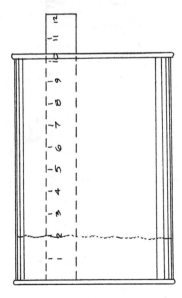

Pour exactly 2 inches (5.08 cm) of water into the can.

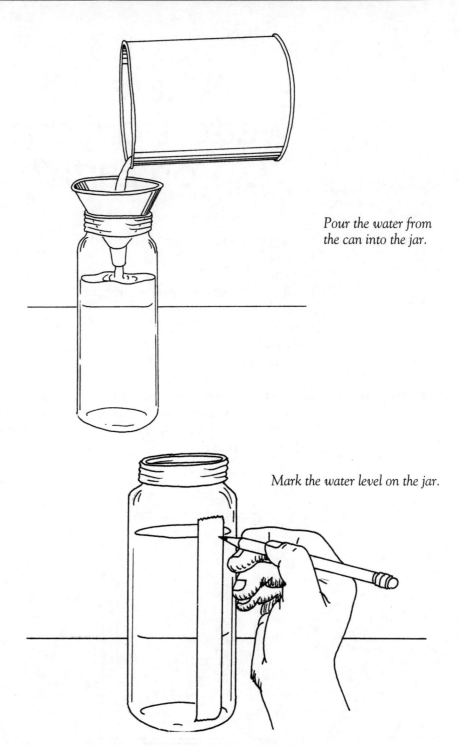

Pour the water from the can into the jar.

Mark the water level on the jar.

The young meteorologist

2. Use the ruler to divide the space below the mark on the jar into 20 equal spaces. This will divide the space into tenths, with each mark representing one-tenth of an inch of rain.

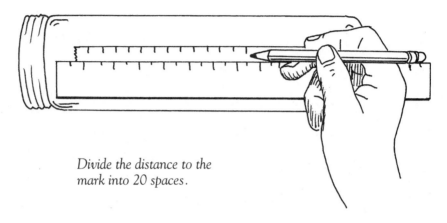

Divide the distance to the mark into 20 spaces.

3. Just before a rain, place the can in an open area away from trees and buildings. Record the time the rain starts and stops, or over a 24-hour period.
4. Use the funnel to pour the collected rainwater into the jar. Use the marks on the jar to measure the amount of rain that fell over the time period you recorded. Use the map to find Cherrapunji, India, in South Asia. Now find Arica, Chile, on the west coast of South America on the map.

results ☆ The experiment measured the rate of rainfall over the time period you recorded. The locations you found on the map are the wettest and driest places on Earth. Cherrapunji, India, has the highest recorded amount of rainfall for one year, 1,042 inches (26,461 mm). Arica, Chile, recorded the lowest yearly rainfall of 0.03 inches (0.8 mm).

further studies ☆ Look up your state in an encyclopedia to find its annual rainfall. How does it compare with India and Chile? What effect does the average precipitation have on agriculture and manufacturing? Do rain forests have any effect on climate around the world?

did you know? ☆ ❏ That Calama, Chile, had no rain for 400 years until 1972 when it had a flood.
❏ That it rains almost every day in the rain forests of South America.

Where is the wettest/driest place on Earth?

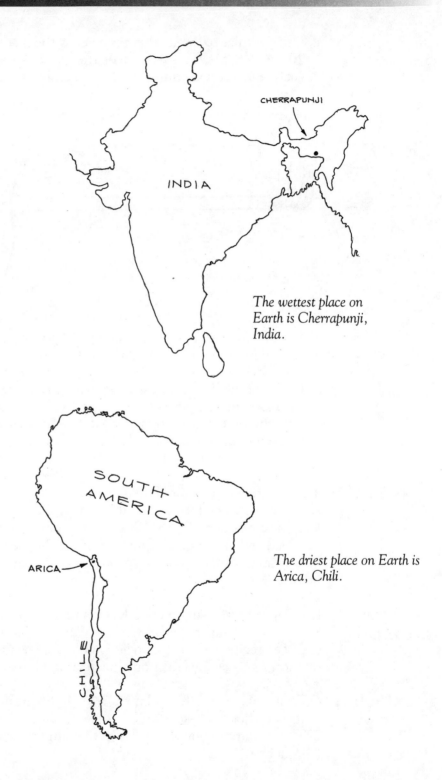

The wettest place on Earth is Cherrapunji, India.

The driest place on Earth is Arica, Chili.

24
Where is the windiest place on Earth?

materials ☆
- ❒ Two thermometers
- ❒ Large jar of water
- ❒ Large jar of dirt

procedure ☆
1. Place a thermometer in each jar and place both jars in the sunlight. Monitor the temperatures of each jar.

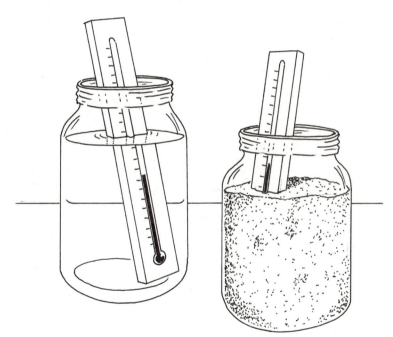

Compare the temperature changes between the water and the soil.

2. Place both jars in a cool, shady place. Monitor the temperatures.

results ☆ In the sunlight, the temperature in the dirt will start to rise first because soil takes in heat quicker than water. In the shade, the soil will lose heat faster than the water. Near oceans during the day, the ground warms the air causing it to rise, allowing the cooler air from the sea to come in. At night, the reverse happens. The ground is cooler than the seawater, so the air over the water rises while the air over the ground flows out. Temperature is one of the biggest factors in creating winds. At the equator, the warm air rises creating a low pressure area. Cold air over the poles sink creating high pressures. Air flows from high- to low-pressure areas, creating winds. The winds do not flow in a straight line toward the equator, but curve sideways because of the rotation of the Earth on its axis. The George V Coast in Antarctica is the windiest place on Earth with gale winds of 200 mph (320 kph). The highest wind gusts of 231 mph (371 kph) were recorded at a weather station on Mount Washington, New Hampshire.

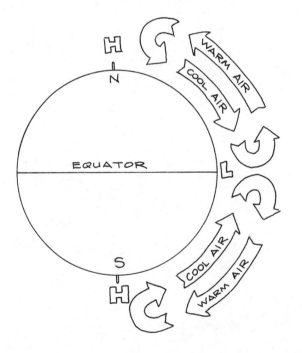

The difference in temperatures creates winds.

The young meteorologist

The highest winds on Earth occur along the southern coast of Antarctica.

The highest wind gust was recorded at a mountain weather station in New Hampshire.

Where is the windiest place on Earth?

further studies ☆ What is wind chill and how does it affect people? Does wind chill affect plants? How many ways can you think of to use wind power? Could modern technology be used to design better sails for ships?

did you know? ☆
- ❐ That "brickfielder" is an Australian term for a very hot summer wind known to blow sand and dust all across the country.
- ❐ That modern windmills use blades that can be more than 295 feet (90 meters) in diameter.

Part 5
The young biologist

Biology, one of the oldest sciences, studies the origin, physical characteristics, habits, and life processes of plants, animals and other living things. Early Egyptians used herbs for medicine and embalming. Leonardo da Vinci and Michelangelo studied biology in their pursuit of art. In the seventeenth century, William Harvey studied the human circulatory system. Louis Pasteur developed heat-sterilization, and Alexander Fleming discovered penicillin in 1928. As knowledge grew and instruments and equipment became more advanced, biologists were able to probe deeper into the mysterious world of living things. Today, biologists are making exciting discoveries that grow better plants, keep us healthier, and even let us live more comfortable lives.

25
Where do plants come from?

materials ☆
- ☐ A few large dried beans
- ☐ Bowl
- ☐ Water
- ☐ Magnifying glass

procedure ☆
1. Place the beans (seeds) in the bowl and cover them with water. Let them soak overnight.

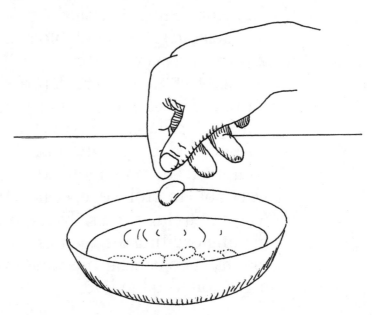

Soak the seeds in a bowl of water.

2. Find a place on one of the seeds where you can spread it open with your thumbnails. Examine the inside of the seed with a magnifying glass.

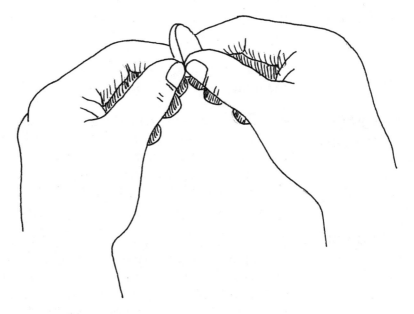

Separate the two halves of one seed.

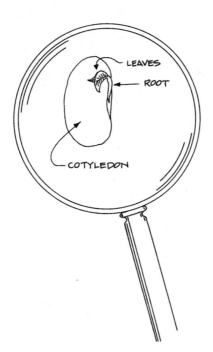

Examine the inside of the seed.

Where do plants come from?

results ☆ You should notice the outer covering of the seed. This is the seed coat. Inside the seed coat is a large area known as the *cotyledon*. This is the food supply for the young plant. Closer examination will show tiny leaves and even the part that will become the root and stem of the future plant.

further studies ☆ Place a couple of the remaining seeds in moist paper towels. Put them in a glass in a warm sunny place and see if they sprout. Put the remaining seeds in moist dirt to see if they grow. Try planting seeds from an orange or a grapefruit. Cut the top off a carrot and place it in a saucer of water. Set the saucer in sunlight. Do you think it will grow?

did you know? ☆
- ❒ That seeds come in a variety of sizes—from tiny dust-like seeds of orchids to much larger seeds, such as coconuts.
- ❒ That some seeds will still germinate even after hundreds of years.
- ❒ That tiny seeds can produce redwood trees more than 300 feet tall and the bristlecone pine tree that lives more than 4,000 years.

26
Where can you see how a plant eats?

materials ☆
- ☐ Jar
- ☐ Red food coloring
- ☐ Water
- ☐ Knife
- ☐ Tweezers
- ☐ Fresh stalk of celery with leaves

procedure ☆

1. Fill the jar about half full of water and add a few drops of food coloring. Have an adult cut about 1 inch (2.5 cm) from the bottom of the stalk and stand it in the jar of red water. Let the jar set until the next day then examine the celery. Has it changed color?

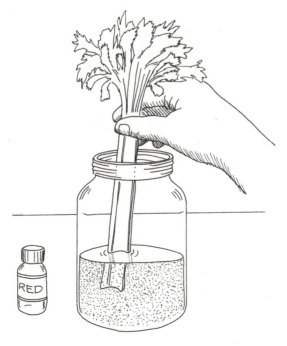

Place a stalk of celery in colored water.

results ☆ Plants take in food through tiny, thread-like parts called *root hairs*. Water and dissolved minerals pass through the thin walls of the root hairs, then up through tubes to the leaves. Your celery did not have root hairs, but you should see red color in the stalk and the leaves. This is how the plant eats.

further studies ☆ Cut across the stalk and examine the tubes that carry the food. Separate the stalk with tweezers and follow the red streaks up to the leaves. What part do the leaves play in plant growth? What is *photosynthesis*? Are all plants green?

did you know? ☆ ❏ That the oxygen in our atmosphere is there only because of green plants.
❏ That plants tend to grow continuously throughout their lives.

27
Where can you see a banana ripen a tomato?

materials ☆
- ☐ Banana (one that still has a little green on the tip)
- ☐ Two small green tomatoes
- ☐ Two jars with lids

procedure ☆
1. Put the banana and a tomato in one jar and screw on the lid. Put the other tomato in a jar by itself and screw on the lid.
2. Place both jars in a dark place at room temperature and let them set for a day. What happened to the tomatoes?

Monitor the changes in each tomato.

results ☆ The tomato in the jar with the banana will ripen but the tomato by itself will still be green. When fruit ripens, substances inside, called *enzymes*, bring about chemical changes.

Enzymes break down the cells, soften the fruit, and change the acid in the fruit to sugar. The green color fades as the chlorophyll collapses, causing the fruit to give off a gas called *ethylene*. A ripening banana is a good source of ethylene. The ethylene gas from the banana caused the tomato in that jar to ripen. Farmers pick green fruit, then just before taking it to market, spray it with an ethylene compound to ripen it.

further studies ☆ Remove the green tomato from the jar and see how long it takes to ripen. Compare the ripening time of the two tomatoes. Do farmers use other compounds to slow the ripening of fruits? How does this process affect the taste of fruits?

did you know? ☆
- That, technically, the tomato is a fruit, but it is generally thought of as a vegetable.
- That the bananas we eat have been cultivated without pollination, are seedless, and cannot reproduce.

28
Where can you see a tadpole turn into a frog?

materials ☆
- ❏ Aquarium
- ❏ Sand
- ❏ Water plants
- ❏ Algae-covered rocks from pond
- ❏ Water
- ❏ A few tadpoles
- ❏ Bits of hard-boiled egg
- ❏ Magnifying glass

procedure ☆

1. Put the aquarium in a place out of direct sunlight. Warm water will kill the tadpoles. Build a small beach of sand at one end of the aquarium. Add a few inches of water, a couple of water plants, and a few algae-covered rocks.

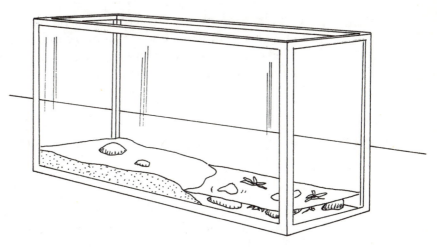

Build a small beach of sand, then add water plants and algae-covered rocks.

2. Drop in about six or eight tadpoles and monitor any changes for the next few months.

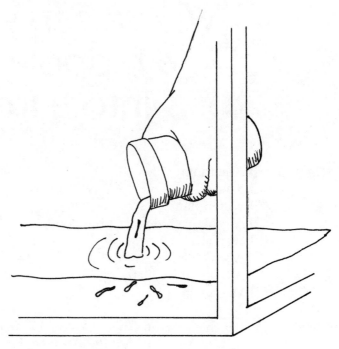

Add a few tadpoles.

results ☆ Tadpoles feed on tiny algae that grow on underwater rocks and the stems of water plants. Older tadpoles can be fed bits of a hard-boiled egg. The tadpoles will slowly grow legs, lose their tails, and become adult frogs. The complete change can take a few months, or in the case of the bullfrog species, up to two years.

further studies ☆ What is *metamorphosis*? Can you think of other species that go through such radical changes from egg to adult? Are frogs cold-blooded?

did you know? ☆ ❏ That frogs live on every continent except Antarctica.
❏ That desert-living toads stay buried in the sand most of the year and come out only when rain forms puddles.
❏ That the tadpoles of desert-living toads must change into adults before the puddle dries up.

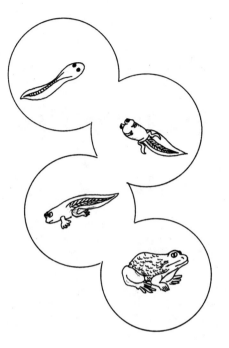

Tadpoles slowly grow into frogs.

Where can you see a tadpole turn into a frog?

29
Where do frogs go in winter?

materials ☆
- ☐ Aquarium
- ☐ Water
- ☐ Dirt
- ☐ Wire screen
- ☐ Frog

procedure ☆
1. Construct a semiaquatic aquarium by building a mud bank leading up to a small hill of dirt at one end and a small pool of water at the other end. Try to make a natural surrounding.
2. Place the frog in the aquarium and feed it live insects such as small bugs and flies. Notice how the frog breathes by watching the pulsing of its throat.
3. In the fall, place the aquarium outside in the cold for several hours. If you have mild winters, scatter ice cubes inside the aquarium. Watch the frog's throat to see any changes in its breathing.

results ☆ At first, you should see rapid pulsing in the frog's throat indicating normal breathing. But as colder temperatures approach, its breathing slows and the frog appears sluggish. Soon, it will burrow into the mud bank and go into hibernation, a long sleep until the warmth of spring wakes it up. Before hibernation, animals normally eat large amounts of food to build up body fat that will sustain them during the winter. In hibernation, the animal's heart rate and breathing are reduced to a minimum to conserve energy and lower its need for food.

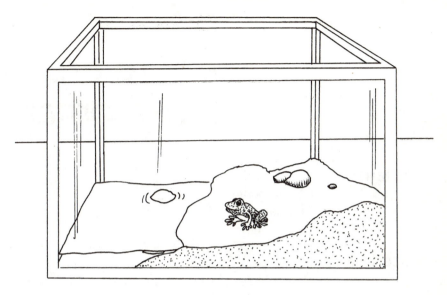

Frogs go into hibernation at the approach of cold weather.

further studies ☆ Frogs are cold-blooded animals. What warm-blooded animals can you think of that also hibernate? Do some bats hibernate? What is an animal's metabolism? Can bear cubs be born during hibernation? Could a form of hibernation be useful in space travel?

did you know? ☆
❏ That some frogs can survive even when 65 percent of their body water has turned to ice.
❏ That some species of African frogs can grow to about 3 feet (1 m) in length.

30
Where are our sense organs?

materials ☆
- ❏ A friend
- ❏ Two pencils
- ❏ Small slices of raw potato and apple
- ❏ Two spoons

procedure ☆
1. Ask your friend to hold the pencil pointing straight up, about 2 feet (60 cm) in front of your eyes. With one eye closed, slowly bring your arm around and place one finger on the point of the pencil. Repeat the step with both eyes open.

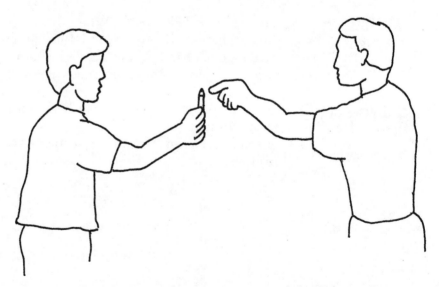

With one eye closed, touch the point of the pencil.

2. Close your eyes and hold your nose, then ask your friend to hand you a slice of potato or apple. Taste the food and try to identify it.

The young biologist

Taste a potato or an apple.

3. Ask your friend to close his or her eyes and place his or her arm across a table, palm up. Hold two pencils side-by-side and touch your friend's finger. Ask your friend to guess how far apart the pencils are. Touch the finger with the pencils farther apart. Repeat the tests on the skin of the upper arm.

Touch the pencils to one finger.

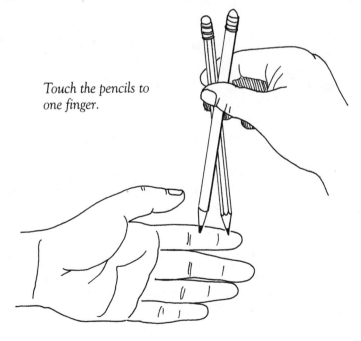

Where are our sense organs?

4. Ask your friend to walk around you, softly clicking the two spoons together. With your eyes closed and without turning your head, point in the direction of the clicking spoons and estimate the distance.

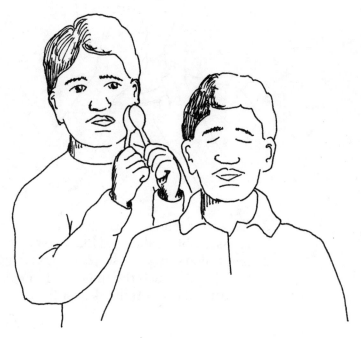

Identify the direction of the sound.

results ☆ All higher animals have sets of sense organs connected to a central nervous system. Receptors in the sense organs transmit signals to the brain in response to various sensations detected by the sense organs. Sense organs are divided into three classes. One class, (*exteroceptive*) provide information relating to sight, hearing, taste, smell, touch, temperature, pain, and pressure. These organs are found on the exterior of the body and include the eyes, ears, nose, tongue, and skin. These are the organs involved in this experiment.

Other sense organs (*interoceptors*) are located mostly in the internal organs of the body. These sense organs provide information relating to sensations such as pain, hunger, thirst, and fatigue. The other class of sense organs (*proprioceptive*) provide information relating to the movement and position of body parts. These organs are located in the muscles, tendons, and joints, as well as in the balance organs in the ear.

It was easier to touch the point of the pencil with both eyes open because with only one eye, you don't have normal depth perception. We use *stereoscopic* vision to determine depths.

It was difficult to tell the difference between the potato and the apple because our sense of taste depends heavily on our sense of smell and sight.

Your friend should be able to guess the distance between the two pencils when you touched them to the finger, but when the pencils touched the upper arm, your friend's guess was probably way off. The nerve endings in our fingers are closer together than in other parts of our body.

You probably were very close at guessing the location of the clicking spoons. Your brain was able to measure the difference between the time it takes sound to reach one ear and then the other. This provides you with the direction and the distance of the source of the sound.

further studies ☆ Look at a distant object. Close one eye then the other. Does the object move to one side then line up again? Does this mean one eye is dominate just like we are right handed or left handed? Try blindfold taste tests of different kinds of soda pop. Can our taste buds be easily fooled? Do internal sense organs tell us when to breath?

did you know? ☆
❏ That birds have the best vision. Their eyes are sometimes bigger than their brains.
❏ That normally women have a better sense of smell than men.

Part 6
The young physicist

Physics is a fascinating science that studies the natural world we live in as well as the unexplored distances of space. We practice the laws of physics every day; when a baseball is pitched, struck, and then flies out to center field, Isaac Newton's laws of motion and gravity are demonstrated; a transformer powering a model electric train proves James Clerk Maxwell's laws of electromagnetism; and when you ride a bicycle, the gyroscopic rotation of the wheels keeps you traveling upright. Physics explains many things such as where thunder comes from, what gravity is, and what makes smoke rise. Drawing on the experiences of early pioneers and recent discoveries, today's physicist has the equipment and technology to explore even deeper into the mysteries of the natural world.

31
Where can you see air pressure make heat?

materials ☆
- ☐ Tire pump
- ☐ Flat tire or basketball

procedure ☆
1. Operate the pump for several minutes of hard pumping to inflate the tire. Feel the lower part of the barrel of the pump.

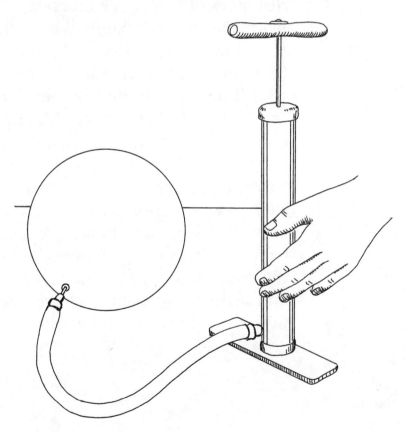

Compressing air generates heat.

results ☆ The bottom part of the pump should be very warm. The heat was produced by friction. It was heated because the molecules making up the air were pressed closer together, and were forced to rub against each other more than normal. There is also some added heat caused by the friction of the piston rubbing against the inside of the barrel.

further studies ☆ Press your thumb against the opening of the pump tube and operate the pump a couple of times. Does your thumb get hot? Does compression generate heat? Place a nail on a piece of heavy iron. Remembering to wear goggles, flatten a place on the nail by striking it several times with a hammer, not too hard, just enough to make a dent in the nail. Does the nail get hot? Were molecules in the nail pressed together?

did you know? ☆
❏ That heat is a form of energy and that cold is the absence of heat. At absolute zero, minus 273.15 degrees C, all molecular motion stops. So far, it has been impossible to attain absolute zero, so everything has some amount of heat.
❏ That heat normally flows only in one direction, from hot objects to cooler ones.

Where can you see air pressure make heat?

32
Where can you see a candle burn steel?

materials ☆
- ☐ Candle and matches
- ☐ Steel wool
- ☐ Pliers

procedure ☆

1. Separate a small piece of steel wool from the larger roll and pull it apart to form a very loose wad.
2. Ask an adult to help you. Wearing gloves and goggles, light the candle and use the pliers to hold the loose wad in the flame.

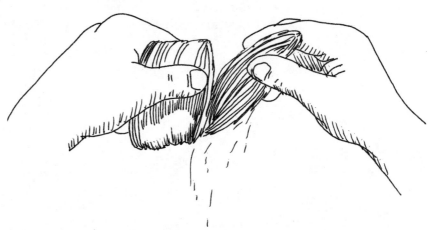

Pull a small piece of steel wool from the roll.

results ☆ The steel wool will burn. Oxygen, which is necessary for burning, is able to surround the thin strands of steel wool. These strands have very little mass to conduct the heat away and easily burn.

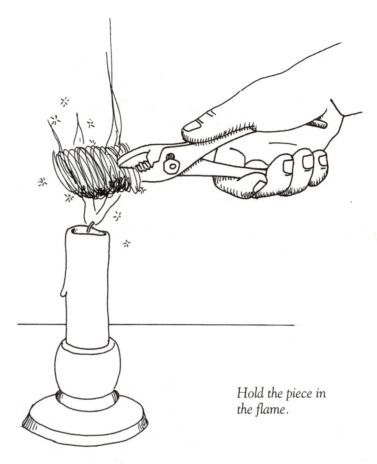

Hold the piece in the flame.

further studies   Hold the steel end of a screwdriver in the candle flame. Would you expect it to burn? Does the metal in the screwdriver have more mass than the steel wool? Why is kindling used to start a fire? What is kindling temperature?

did you know?
- ❒ That the fatal tragedy that struck the Apollo crew in 1967 was caused when an electrical spark ignited fire-resistant plastics that became flammable because of pure oxygen in the cabin.
- ❒ That oxyacetylene torches used to cut and weld metals have flames that can reach 6,000 degrees F (3,315.5 degrees C).
- ❒ That certain things like hay, coal, and oily rags can ignite by themselves by a process of slow oxidation called spontaneous combustion.

Where can you see a candle burn steel?

33
Where can you see a needle floating in air?

materials ☆
- ☐ Sewing needle with about 12 inches (30.48 cm) of thread
- ☐ Horseshoe magnet

procedure ☆

1. Magnetize the needle by stroking it across one end of the magnet.

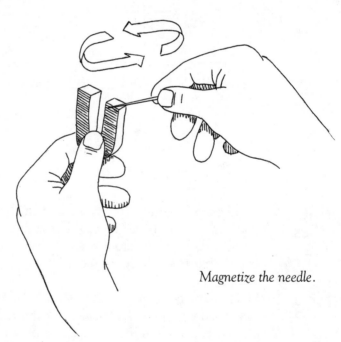

Magnetize the needle.

2. Carefully lift the needle by the thread and, still holding the thread, slowly try to lay the needle across the other pole of the magnet.

The young physicist

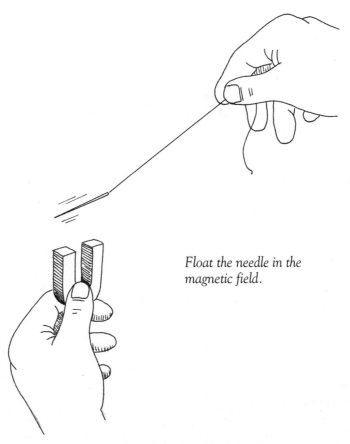

Float the needle in the magnetic field.

results ☆ When you lifted the needle by the thread, it continued to point to the first pole of the magnet. As you moved the needle over the other pole, it became suspended by the balance between the attraction of one pole and the repulsion of the other.

further studies ☆ Magnetize two needles by rubbing one in just one direction across a magnet. Magnetize the other needle by rubbing it back and forth across the magnet. Did both needles become magnetized? Did the direction of stroking the needles make any difference? Could you separate the poles of a bar magnet if you cut the magnet in half?

did you know? ☆ ❐ That some species (birds, stingrays, whales, etc.) can sense the lines of force of the Earth's magnetic field and use them for navigation.
❐ That the Earth is a huge magnet and was described as such in 1600 by William Gilbert, the physician to Queen Elizabeth I.

34
Where can you see a hammer defy gravity?

materials ☆
- String about 12 inches (30.48 cm) long
- Ruler
- Hammer
- Table

procedure ☆
1. Tie the string to form a loop about 5 inches (12.7 cm) across. Slip the loop about two-thirds of the way down the ruler, and about halfway down the handle of the hammer.
2. Place the tip of the ruler on the edge of the table. The head of the hammer should be slightly under the table with the end of the handle about one-third of the way down the ruler.
3. Release the ruler and hammer.

results ☆ It might take a little adjusting, but the hammer should balance with the tip of the ruler on the edge of the table. It doesn't look like it, but the center of gravity has been positioned directly below

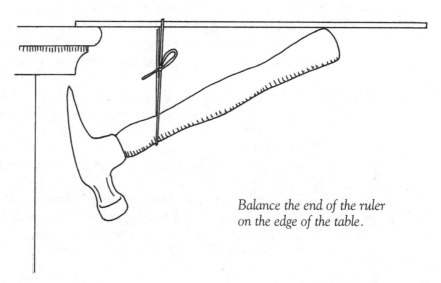

Balance the end of the ruler on the edge of the table.

the edge of the table. The center of gravity is the one point where all of the mass of an object seems to exist. If an upward force is exerted at the center of gravity, equal to the object's weight, the object will be balanced.

further studies ☆ Stand the ruler in the palm of your hand and try to make it balance. Now try a yardstick or a broom. Which is easier to balance? Does a greater distance from the force (your hand) to the mass of the object make it easier to balance? Why do circus performers use long poles, that bend down slightly, to balance on a tightrope?

did you know? ☆
- ❏ That the greater the distance from the Earth, the less gravity would affect you. For example if you weigh 100 pounds on Earth, at 8,000 miles in space you would only weigh about 25 pounds.
- ❏ That you would weigh slightly more at the poles than at the Equator.

35
Where can you see a diving medicine dropper?

materials ☆
- ☐ Plastic bottle with lid
- ☐ Water
- ☐ Medicine dropper

procedure ☆
1. Fill the plastic bottle three-quarters full with water. Fill the medicine dropper with water and float the medicine dropper in the bottle. You might have to adjust the amount of water in the medicine dropper so that the rubber end floats a little above the surface. Screw on the lid of the plastic bottle.

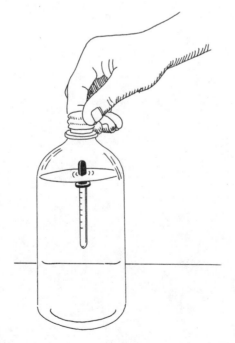

Float the medicine dropper in the bottle.

2. Squeeze the bottle with your hands, then release the pressure. What happens to the medicine dropper?

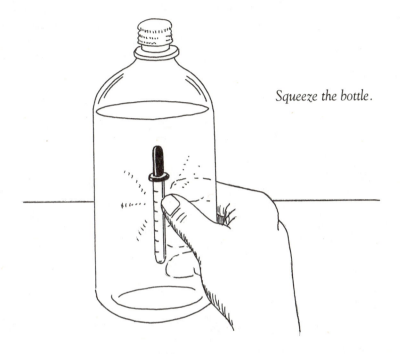

Squeeze the bottle.

results ☆ The medicine dropper floated at the surface until the bottle was squeezed. Then it sank toward the bottom. When the pressure was released, it floated back to the surface. Water is hard to compress, while air can be compressed easily. When the plastic bottle was squeezed, the water was not compressed, but the air inside the medicine dropper was. This allows more water to enter the medicine dropper and it sinks. When the pressure was released, the compressed air in the medicine dropper pushes some of the water out and the medicine dropper rises. This is the basic principle for controlling submersible crafts.

further studies ☆ Try to make the medicine dropper float at different levels by varying the pressure on the bottle. Would air make better shock absorbers than water? Would a hydraulic jack work well with air instead of fluid?

Where can you see a diving medicine dropper?

did you ☆
know?
- That the submersible craft *Trieste*, in 1960, descended 35,800 feet (10,912 m) into the Marianas Trench near the Philippines.
- That dolphins can dive to more than 1,000 feet (300 m), but the Weddell seal can descend to depths of 2,000 feet (600 m) and stay submerged for more than one hour.

36
Where can you see a Ping-Pong ball floating in the air?

materials ☆
- ❐ Hair dryer or tank-type vacuum cleaner
- ❐ Ping-Pong ball

procedure ☆

1. Turn the hair dryer on high without heat, or ask an adult to reverse the hose connections on the vacuum cleaner and turn it on. Aim the nozzle straight up and float the Ping-Pong ball on the column of air.

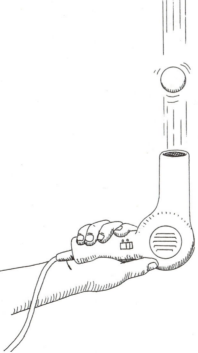

Float the Ping-Pong ball in the air.

2. Move the ball slightly to one side of the column of air then release it. Does the ball fall?

results ☆ The Ping-Pong ball floated on the column of air. The air flows faster in the center of the column than near the outside. This faster flow of air creates a low pressure area in the center of the column. This low pressure keeps the ball from drifting out of the column of air.

further studies ☆ Tilt the column of air to see how steep an angle can be achieved before the ball falls. Tape a cardboard tube to the hair dryer, place a funnel in the tube and place the Ping-Pong ball in the funnel. Turn on the dryer. Does the ball float? What happens if you point the dryer down? Blow a stream of air over the top of a sheet of paper. What happens to the paper?

did you know? ☆
- ❏ That the faster air, or any fluid, moves, the more the pressure drops. This effect is known as Bernoulli's Law, named after Daniel Bernoulli, who discovered the principle in 1738.
- ❏ That the curved surface of an airplane wing moving through the air creates a pressure low enough to lift the airplane.

37
Where can you see air lifting water?

materials
- Drinking straw
- Hobby knife
- Glass of water

procedure

1. Ask an adult to carefully cut a slit crossways in the straw about two inches (5.08 cm) from one end. Be sure not to cut all the way through the straw.

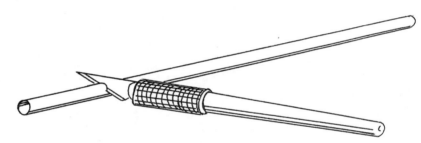

Cut a slit in the straw.

2. Rub your finger down the longer part of the straw to flatten it and bend the straw at the cut. Do not flatten the short part of the straw.
3. Put the short end of the straw in the glass of water so that the bend is just above the surface and at the far edge of the glass. Blow hard through the long part of the straw.

results A mist of vapor will spray from the opening in the bend. The jet of air you blew through the long part of the straw lowered the air pressure in the short part sticking in the water. The normal air pressure pressing down on the surface of the water forces it up through the short end of the straw where it is blown into a mist.

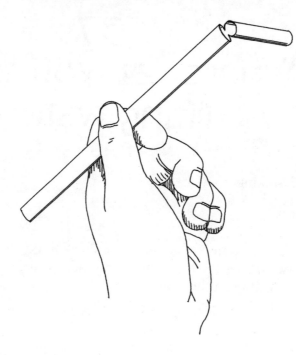

Bend the straw at the cut.

Blow through the straw.

further studies ☆ Could air flowing through a tube be used to siphon fluids? Would such a system be safe for transferring hazardous chemicals? Could spray painters use the same principle to apply paint?

did you know? ☆
- ❐ That a dentist uses the same principle to keep fluids from collecting in your mouth while work is being done.
- ❐ That a carburetor in a car sprays fuel when a jet of air passes over a small tube.

38
Where can you see sound patterns?

materials ☆
- ☐ Large baking pan
- ☐ Medicine dropper
- ☐ Water

procedure ☆
1. Place the pan on a level surface and fill it with about one inch (2.54 cm) of water. Let the water stand until it is smooth and flat.
2. Use the medicine dropper to drop a single drop of water into the pan near one end. Notice the pattern of the ripples as they spread across the water.

results ☆ The ripples spread out from the point where the drop hit the surface of the water. When they strike the sides of the pan, the

Ripples in water create patterns similar to sound waves.

ripples bounce back in a new direction. This action continues until the ripples fade away. This is the same pattern sound waves follow. When you speak or make some other sound, sound waves move in a pattern that is similar to the ripples in the water.

further studies ☆ Drop a drop of water in one of the corners and see what kind of pattern appears. Bend a strip of metal like a hack saw blade into a curve and wedge it in the pan. Drop several steady drops of water in front of the curved strip. What kind of pattern appears? Why do you suppose the back of the stages in some theaters and most concert halls are curved?

did you know? ☆
- That sound needs a solid, liquid, or gas to travel through. Sound cannot exist in a vacuum.
- That sound levels are measured in decibels (dB), and normal conversation is at about 60 dB, a rock band at about 110 dB, and about 140 dB can produce pain.

Appendix

Museums for seeing where science happens

Alabama — The Exploreum, 1906 Spring Hill Ave., Mobile. Exhibits relating to physics (optics and energy) and biology (bone manipulation, fresh water and seawater environment).

U.S. Space and Rocket Center, One Tranquility Base, Huntsville. Exhibits include a centrifuge machine and a life-size space shuttle exhibit.

Alaska — The Imaginarium, 725 West 5th Ave., Anchorage. Exhibits relating to biology (polar bears, and marine life), and a planetarium.

University of Alaska Museum, 907 Yukon Dr., Fairbanks. Exhibits of fossils, mammals, birds, and plants.

Arizona — Arizona Museum of Science and Technology, 80 N. Second St., Phoenix. Exhibits relating to biology (reptiles, human body and nutrition) and physics (energy).

Tucson Children's Museum, 200 South Sixth Ave., Tucson. Exhibits relating to biology (models of body parts, a transparent animated human) engineering (modes of transportation, wheels).

Arkansas — Arkansas Museum of Science and History, MacArthur Park, Little Rock. Exhibits relating to biology (bear's habitat, bones and fossils) and meteorology (weather forecasting).

Mid-America Museum, 500 Mid-America Blvd., Hot Springs National Park, Hot Springs. Exhibits relating to physics (optics, magnets, gravity, and electricity) and biology (regional plants and animals).

California Balboa Park Museums, San Diego, Space Theater and Science Center. Several science exhibits including a gravity well.

California Academy of Sciences, Golden Gate Park, San Francisco. Exhibits include a live coral reef, seals, dolphins, a simulated earthquake, and planetarium.

California Museum of Science and Industry, 700 State Dr., Los Angeles. Exhibits relating to pollution, earthquakes, and chemistry.

The Discovery Center, 1944 North Winery Ave., Fresno. Exhibits relating to biology (worm farm, desert tortoise and garden) and physics (holograms, sound, color optics).

Exploratorium, 3601 Lyon St., San Francisco. Hundreds of exhibits relating to a variety of science fields including a miniature tornado, optics, and sound waves.

Lawrence Hall of Science, University of California, Berkeley. Exhibits include animated giant dinosaurs, biology lab, lasers, life-size model of the space shuttle, and a planetarium.

Sacramento Science Center, 3615 Auburn Blvd., Sacramento. Exhibits changed twice a year include biology, engineering, physics, and a planetarium.

The Tech Museum of Innovation, 145 W. San Carlos St., San Jose. High-tech exhibits of space technology, biology, and engineering.

Colorado Denver Museum of Natural History, 2001 Colorado Blvd., Denver. Exhibits of prehistoric fossils, Colorado wildlife, and a planetarium.

Discovery Center, University Mall, Fort Collins. Science exhibits of pulleys, air cannon, and lasers.

Connecticut Discovery Museum, 4450 Park Ave., Bridgeport. Exhibits that include simulated space trip, experiments of Michael Faraday and Alexander Bell.

Thames Science Center, Gallows Lane, New London. Exhibits include biology (water and the environment) and engineering (forces of water).

Delaware Delaware Museum of Natural History, 4840 Kennett Pike, Wilmington. Exhibits of sea life, birds, fossils, and sea shells.

Florida The Discovery Center, 231 South West Second St., Ft. Lauderdale. Exhibits on biology (sea life, insects, human skeleton, animal skulls) and physics (hands-on electric generator).

Museum of Science and Industry, 4801 East Fowler Ave., Tampa. Exhibits on physics (huge pinball machine demonstrating motion and energy) and meteorology (inside of a hurricane).

Museum of Science and Space Transit Planetarium, 3280 South Miami Ave., Miami. Exhibits on biology (human body and muscles) physics (lasers, sound, magnetics, optics), and a planetarium.

Orlando Science Center, 810 East Rollins St., Orlando. Exhibits on biology (live snakes, alligators), meteorology (miniature tornado), engineering (canals and dams), and a planetarium.

South Florida Science Museum and Planetarium, 4801 Dreher Trail North, West Palm Beach. Exhibits include a high-voltage generator, robots, sea life, and daily planetarium shows.

Georgia The Museum of Arts and Sciences, 4182 Forsyth Rd., Macon. Changes its exhibits relating to art and science.

Hawaii Bishop Museum, 1525 Bernice St., Honolulu. Exhibits of birds, plants, and planetarium showing stars used for navigation.

Idaho The Discovery Center of Idaho, 131 Myrtle St., Boise. Exhibits of optics, a gyroscope, and energy.

Idaho Museum of Natural History, Idaho State University, Pocatello. Exhibits include animated dinosaurs, fossils, plants, and minerals.

Illinois Decatur Area Children Museum, Rock Springs Environmental Center, 1495 Brozio Lane, Decatur. Exhibits include the Bernoulli Principle, and sound.

Discovery Center Museum, 711 N. Main St., Rockford. Exhibits on magnetism, electricity, machines, optics, health, and a planetarium.

Exploration Station, 396 N., Kennedy Dr., Bradey. Exhibits of early Dutch scientists, operating windmill, and earth science.

Museum of Science and Industry, 57th and Lake Shore Dr., Chicago. Exhibits of simulated space shuttle launch, human brain, and inside a coal mine.

SciTech, 18 West Benton St., Aurora. Exhibits of magnetism, optics, pendulum, and Sun.

Indiana The Children's Museum, 3000 N. Meridian St., Indianapolis. Exhibits on environmental sciences, live animals, electricity, optics, sound, and engineering.

Evansville Museum of Arts and Science, 411 South East Riverside Dr., Evansville. Exhibits of minerals, rocks, and planetarium.

Iowa Science Center of Iowa, 4500 Grand Ave., Des Moines. Exhibits on physics, lasers, live animals, and a planetarium.

Science Station, 427 1st South East, Cedar Rapids. Many science exhibits including operating a model hot air balloon.

Kansas Kansas Learning Center for Health, 505 Main St., Halstead. Biology exhibits of human body, most of the parts, and how they work.

Omnisphere and Science Center, 220 South Main St., Wichita. Exhibits covering gravity, electricity, optics, astronomy (planetarium) and chemistry.

Louisiana Louisiana Nature and Science Center, 11000 Lake Forest Blvd., New Orleans. Exhibits relating to plants, small animals, and insects.

Maryland Maryland Science Center, 601 Light St., Inner Harbor, Baltimore. Exhibits on engineering, energy, and a planetarium.

Maryland Academy of Sciences, 601 Light St., Baltimore. Exhibits of live animals, aquariums, rocks, energy, and planetarium.

Massachusetts Children's Museum, 300 Congress St., Boston. Exhibits include human skeleton, laws of physics using golf balls, and people of different cultures.

The Discovery Museum, 177 Main St., Acton. Huge exhibit demonstrating chain reaction, a gravity well, and a do-it-yourself inventor's lab.

Springfield Science Museum, 236 State St., Springfield. Exhibits include human life, dinosaurs, an underwater look at turtles, and a planetarium.

Michigan Ann Arbor Hands-On Museum, 219 East Huron St., Ann Arbor. Exhibits include pulleys, robots, optics, a hologram, motor skills, live birds, and African frogs.

Cranbrook Institute of Science, 500 Lone Pine Rd., Bloomfield Hills. Do-it-yourself physics lab, lasers, live animals, animal bones, and a planetarium.

Impression 5 Science Museum, 200 Museum Dr., Lansing. Do-it-yourself chemistry lab, robots, electronics, and generating electricity.

Minnesota Science Museum of Minnesota, 300 East 10th St., St. Paul. Exhibits of light, color, different types of waves, fossils, and prehistoric life forms.

Mississippi	Mississippi Museum of Natural Science, 111 North Jefferson St., Jackson. Biology exhibits include bears, skulls, fossils, and river ecosystems.
Missouri	St. Louis Science Center, 5050 Oakland Ave., St. Louis. Exhibits of fiber optics, virtual reality, building with Styrofoam blocks, and a planetarium.
Montana	Museum of the Rockies, Montana State University, 600 W. Kagy, Bozeman. Exhibits of dinosaurs and other prehistoric life.
Nebraska	Encounter Center, University State Museum, 307 Morril Hall, Lincoln. Exhibits of reptiles, fish, insects, fossils, and replica of a dodo bird and Bengal tiger.
	Lincoln Children's Museum, 121 South 13th St., Lincoln Square, Lincoln. Exhibits include a spacecraft ride, and inside the human body.
Nevada	Lied Discovery Children's Museum, 833 Las Vegas Blvd., North, Las Vegas. Many exhibits including a bubble machine.
New Jersey	Liberty Science Center, 251 Phillip St., Jersey City. Many exhibits including an inventor's lab, and human health and the environment.
	New Jersey Children's Museum, 599 Industrial Ave., Paramus. Exhibits that include a helicopter, flight simulator, simple machines, and exhibits on Thomas Edison and Albert Einstein.
New Mexico	Santa Fe Children's Museum, 1050 Old Pecos Trail, Santa Fe. Exhibits about magnets, water tables, and wildlife.
New York	International Museum of Photography at George Eastman House, 900 East Ave., Rochester. Exhibits of inside a camera, darkroom techniques, pin hole camera, history of cameras, and early photographs.
	Museum of Science and Technology, Armory Square, Syracuse. Exhibits include a momentum machine, hologram, giant lens,

human lungs, waves (sound, light, and water), chemistry lab, and a planetarium.

New York Hall of Science, 4701 111th St., Flushing Meadows, Corona Park. Exhibits on logic and mechanics, light and color, and engineering principles.

Roberson Museum and Science Center, 30 Front St., Binghamton. Exhibits include a giant kaleidoscope, image and light, a space exhibit, and a planetarium.

Science Discovery Center, State University College, Oneonta. Exhibits on sound, sound waves, magnetics, and optics.

Sciencenter, Ithaca. Engineering exhibits about water flows and pressure, the human body, measurements, and a view from inside a camera.

North Carolina Discovery Place/Nature Museum, 301 North Tryon St., Charlotte. Exhibits include an ocean aquarium, a tropical rain forest, and subzero temperatures.

Nature Science Center, Museum Dr., Winston-Salem. Exhibits of animals, sea life, sound waves, light, electrostatic generator, and a planetarium.

North Carolina Museum of Life and Science, 433 Murray Ave., Durham. Exhibits on muscles and bones, a bear's skeleton, and shells and fossils from local beaches.

Ohio Cleveland Children's Museum, 10730 Euclid Ave., Cleveland. Physics exhibits using water, boat and bridge building.

Ohio's Center of Science and Industry, 280 East Broad St., Columbus. Many science exhibits including human lungs and heart.

Oklahoma Omniplex Science Museum, 2100 N.E. 52nd St., Oklahoma City. Exhibits include a mirror maze, Earth's and Moon's gravity, high-voltage electricity, weather, tropical gardens, and planetarium.

Oregon — Oregon Museum of Science and Industry, 1945 S.E. Water Ave., Portland. Exhibits include space capsule, wind tunnel, computers, lasers, fiber optics, and a planetarium.

Willamette Science and Technology Center, 2300 Leo Harris Pkwy., Eugene. Exhibits include a solar telescope, shadow wall, alternating science exhibits, and nearby planetarium.

Pennsylvania — Academy of Natural Sciences, 1900 Benjamin Franklin Pkwy., Philadelphia. Exhibits include fish, frogs, turtles, whale bones, dinosaurs, and trip inside the earth.

Carnegie Science Center, One Allegheny Ave., Pittsburgh. Exhibits include World War II submarine, coral reef, lasers, water table, and a planetarium.

The Franklin Institute, Benjamin Franklin Pkwy 20th St., Philadelphia. Exhibits of static electricity, electricity, communications, ship designing, flight simulator, a rain forest, and a planetarium.

Museum of Scientific Discovery, Third and Walnut Sts., Harrisburg. Exhibits on aviation, bridge building, optics, and math.

South Dakota — Discovery Center and Aquarium, 805 W. Sioux, Pierre. Exhibits include a gyro chair, optics, hot air balloon, and aquarium with salmon spawning.

Tennessee — Cumberland Science Museum, 800 Ridley Blvd., Nashville. Exhibits include "Kinetic Coaster" demonstrating laws of physics, human development, reflex tests, and model of human brain.

East Tennessee Discovery Center, 516 N. Beamon St., Chilhowee Park, Knoxville. Exhibits of mechanics, light, sound, fossils, fresh and seawater aquariums, and a planetarium.

Texas — Don Harrington Discovery Center, 1200 Streit Dr., Amarillo. Exhibits about electricity, sound, dams and canals, house building, and a planetarium.

Fair Park Museums, Dallas Museum of Natural History, Dallas. Exhibits of fossils, shells, alligators, and live animals.

Fair Park Museums, The Science Place/Southwest Museum of Science and Technology, Fair Park, Dallas. Exhibits of the human body, muscles, skeleton, magnetic fields, levers, water table, and principles of physics.

Houston Museum of Natural Science, 1 Hermann Circle Dr., Houston. Exhibits include a weather station, lasers, mechanics, dinosaurs, space flight with mission control, and a planetarium.

Insights/El Paso Science Center, Inc., Santa Fe and Missouri Sts., El Paso. Exhibits about optics, lasers, sound, solar power, and mechanics.

McAllen International Museum, 1900 Nolana, McAllen. Exhibits on astronomy, meteorology, earth sciences, prehistoric life, and atmospheric and space science.

Science Spectrum, 5035 N. 50th St., Lubbock. Exhibits include live animals, aquarium, optics, gravity maze, and a magnetic pendulum.

Vermont Discovery Museum, 51 Park St., Essex Junction. Exhibits include live animals, reptiles, and a TV weather station.

Virginia Virginia Living Museum, 524 N. Clyde Morris Blvd., Newport News. Exhibits of live animals, waterfowl, seawater life, and stream and river creatures.

Science Museum of Virginia, 2500 West Broad St., Richmond. Exhibits on optics, electricity, electronics, natural crystals, planetarium, and a few miles away, an aviation museum with a flight simulator.

Science Museum of Western Virginia, One Market Sq., Roanoke. Exhibits on earth sciences, human health, biology (owls, bats, and marine animals), a TV weather station, and planetarium.

Washington Pacific Science Center, 200 Second Ave., North, Seattle. Exhibits include life-size mechanical dinosaurs, human muscles and reflexes, and gravity and motion (balancing a bike on a suspended rail).

The Whale Museum, 62 First St., North, Friday Harbor. Exhibits about whales, migration and tracking, and identifying sounds and songs.

Washington, D.C. Capitol Children's Museum, 800 Third St., N.E. Exhibits about communication (prehistoric to satellites) and animation (cartoons).

Smithsonian Institution, Museum of Natural History, on National Mall. Exhibits of fossils (prehistoric shark jaws, dinosaurs), live coral reef, insect zoo, and many skeletons.

Smithsonian Institution, Air and Space Museum, on National Mall. Exhibits include moon rock, airplanes, space shuttle replica.

West Virginia Sunrise Museum, 746 Myrtle Rd., South Hills, Charleston. Exhibits include live animals, gardens, and planetarium.

Wisconsin Discovery World: The Museum of Science, Economics and Technology, 818 W. Wisconsin Ave., Milwaukee. Exhibits of electric power plant, magnetism, lasers, lightning bolts, gyroscope, gravity well, solar system, and artificial heart and human fitness.

Madison Children's Museum, 100 State St., Madison. Exhibits include light, water, and sound waves, engineering and architect (drafting tools, using beams), health and fitness, hospital room with surgery, dairy products, and cows.

Wyoming Wyoming Children's Museum and Nature Center, 710 Garfield St., Laramie. Exhibits about optics, gravity, electricity, sound, live animals, skulls, and a beaver lodge.

Index

A
absolute zero, 103
acids, 53
adsorption, 58
aeronautical engineering, 1
  airlines, first, 2-3, **3**
  airships, 3
  instrument flight first, 8-11, **8, 9, 10**
  parachutes, first jump, 4-7, **4, 5, 6**
air pressure, 61-63, **61, 62**
  Bernoulli effect, 113-114, **113**
  floating objects, 110-112, **110, 111**
  heat caused by air pressure/friction, 102-103, **103**
  venturi effect, 115-117, **115, 116**
airlines, first, 2-3, **3**
airships, 3
alkalis, 53
aluminum, 47-48
Antarctica, coldest place on Earth, 72-74, **72, 73, 74**
antibiotics, 83
Archimedes discovery of scientific laws of levers, 13, 15
Arica, Chile, as driest place on Earth, 77, **78**
astronomy, 25-41
  Babylonian astronomy, 25
  Betelgeuse star, 26, **28**
  Big and Little Dipper, 26-28, **27, 28**
  Cassiopeia constellation, 26, **27**
  Copernicus and Earth-centered solar system, 25
  Draco constellation, 26
  Dubhe, pointer star for Big Dipper, 26, **27**
  Earth-centered solar system, 25
  flatness of Earth at poles, 70
  Galileo and first telescope, 25
  Hubble Space Telescope, 25
  International Date Line, 32-33, **32, 33, 34**
  longitude, 32-33, **32, 33,** 37
  Merak, pointer star for Big Dipper, 26, **27**
  meridians of longitude, 37
  North Star, 26-28, **27, 28,** 30, **30,** 31, **31**
  Orion constellation, 26, **28**
  Palomar Observatory, California, 25
  photographing star tracks, 29-31, **29, 30, 31**
  Polaris, 26-28, **27, 28**
  Prime Meridian, 35-38, **35, 36, 37, 38**
  Rigel star, 26, **28**
  Royal Greenwich Observatory, Prime Meridian, 38
  solar (radiant) energy, 39-41, **39, 40**
  speed of Earth's rotation, 38
  telescopes, 25
  time zones, 33, **33,** 38
atomic theory, 43

B
Babylonian astronomy, 25
barometric measure of air pressure, 63
Benoist Airline, 2
Benoist, Thomas, early airline entrepreneur, 2
Bernoulli effect, 113-114, **113**
Bernoulli, Daniel, 114
Berry, Albert, early parachutist, 7
Betelgeuse star, 26, **28**
Big and Little Dipper, 26-28, **27, 28**
Billings, MT, as greatest temperature change in U.S., 74, **74**
biology, 83-99
  Alexander Fleming and penicillin, 83
  antibiotics, 83
  circulatory system, 83
  cotyledon of seeds, 86
  dolphins deep-dive records, 112
  enzymes and ripening process of fruits, 90
  ethylene gas and ripening process of fruits, 90
  exteroceptive sense organs, 98
  hearing, 98, **98,** 99
  hibernation of frogs, 94-95, **95**
  high blood pressure or hypertension, 54
  interoceptor sense organs, 98
  Louis Pasteur and sterilization techniques, 83
  magnetic fields, navigation by animals, 107
  metamorphosis, tadpole-to-frog, 91-93, **92, 93**
  pasteurization, 83
  penicillin, 83
  photosynthesis in plants, 88
  plants, seeds, germination, 84-86, **84, 85, 86**
  proprioceptive sense organs, 98-99
  ripening process of fruits, 89-90, **89**
  root hairs and plant absorption of water/nutrients, 88
  seeds, age vs. germination, 86
  seeds, size varies with species, 86
  sense organs, 96-99, **96, 97, 98**
  smell, 96, **97,** 99
  sterilization, 83
  tadpole-to-frog metamorphosis, 91-93, **92, 93**
  taste, 96, **97,** 99
  touch, 97, **97,** 99
  Leonardo da Vinci's early study of biology, 83
  vision, 96, **96,** 99
  water/nutrient absorption by celery, 87-88, **87**
  Weddell seal deep-dive records, 112
  William Harvey and blood circulation, 83
Blanchard, Jean Pierre, early parachutist, 7
"brickfielder" winds of Australia, 82
Broadwick, Georgia, first woman parachutist, 7

C
carburetors and venturi effect, 117
Cassiopeia constellation, 26, **27**
Celsius, Anders, 69, 70
Celsius/Fahrenheit temperatures, 68-70, **68, 69**
Channel Tunnel, 18, 19
chemicals producing electricity, 23-24, **23, 24**
chemistry, 43-58
  acids, 53
  adsorption, 58
  alkalis, 53
  aluminum citrates, 47-48, **47**
  atomic theory, 43
  chemicals producing electricity, 23-24, **23, 24**
  Dalton's atomic theory, 43
  digestion and enzyme action, 46
  enzymes and ripening process of fruits, 90
  ethylene gas and ripening process of fruits, 90
  molecular mixture (diffusion), 55-56, **55**
  molecular separation (chromatography), 57-58, **57, 58**
  oxidation with candle and steel wool, 104-105, **104, 105**
  rust (oxidation) formation, 49-52, **49, 50, 51**
  salts, 53-54, **53, 54**
  soap, eggs, enzymes, 44-46, **44, 45**
  solubles, 53
Cherrapunji, India, wettest place on Earth, 77, **78**

chromatography (molecular separation), 57-58, **57**, **58**
circulatory system, 83
citrus to polish aluminum, aluminum citrates, 47-48, **47**
coldest place on Earth, 72-74, **72**, **73**, **74**
computers in meteorology, 59
Copernicus and Earth-centered solar system, 25
cotyledon of seeds, 86

D

Dallol, Ethiopia, hottest average climate on Earth, 73
Dalton, John, discovery of atomic theory, 43
de Saussure, Horace, human-hair hyrometer, 67
decibels to measure sound, 119
dental drill invented, 19
Deutsche Luftschiffahrts Aktien Gesellschaft (DELAG) airline, 2
diffusion (molecular mixing), 55-56, **55**
digestion and enzyme action, 46
dolphins deep-dive records, 112
Doolittle, James, first instrument flight landing, 10, **10**
Draco constellation, 26
drills as simple tools, 16-19, **17**, **18**
Dubhe, pointer star for Big Dipper, 26, **27**

E

Earth as giant magnet, 107
Earth-centered solar system, 25
eggs, soap, enzymes, 44-46, **44**, **45**
electrical engineering, 1
   chemicals producing electricity, 23-24, **23**, **24**
   electricity for heat/light, 20-22, **20**, **21**, **22**
   galvanic action, 23-24, **23**, **24**
   incandescent vs. fluorescent lights, 22
   lightning and electrical discharge, 22
   magnetic fields and electrical forces, 24
electromagnetism, 101
engineering, 1-24
   aeronautical engineering, 1
   airlines, first, 2-3, **3**
   chemicals producing electricity, 23-24, **23**, **24**
   drills as simple tools, 16-19, **17**, **18**
   electrical engineering, 1
   electricity for heat/light, 20-22, **20**, **21**, **22**
   galvanic action, 23-24, **23**, **24**
   instrument flight first, 8-11, **8**, **9**, **10**
   machines, 12-15, **12**, **13**, **14**, **15**
   mechanical engineering, 1

parachutes, first jump, 4-7, **4**, **5**, **6**
enzymes, 44-46, **44**, **45**
   ripening process of fruits, 90
Ethiopia as hottest average climate on Earth, 73
exteroceptive sense organs, 98
eyes (*see* vision), 96

F

Fahrenheit, Gabriel, 69, 70
Fahrenheit/Celsius temperatures, 68-70, **68**, **69**
Fansler, Percival, early airline entrepreneur, 2
Fleming, Alexander, penicillin, 83
fluorescent vs. incandescent lights, 22
force-multiplying levers, 14
friction of air causes heat, 102-103, **103**
frogs
   hibernation, 94-95, **95**
   metamorphosis, 91-93, **92**, **93**
fulcrum-type levers, 13, **13**, **14**

G

Galileo and first telescope, 25
Galvani, Luigi, discovery of galvanic action, 24
galvanic action, 23-24, **23**, **24**
Garnerin, Andre Jacques, early parachutist, 7
George V Coast, Antarctica, windiest place on Earth, 80, **81**
germination of seeds, 84-86, **84**, **85**, **86**
Gilbert, William, Earth's magnetic field, 107
gravity, 101, 108-109, **108**
greatest temperature change in U.S., 74, **74**
Greenwood, John, dental-drill inventor, 19
gyroscopic rotation, 101

H

Harvey, William, blood circulation, 83
hearing, 98, **98**, 99
Hegenberger, Albert, first instrument takeoff and landing, 11
herbal medicine, 83
hibernation of frogs, 94-95, **95**
high blood pressure or hypertension, 54
hottest place on Earth, 71-74, **71**, **72**
Hubble Space Telescope, 25
human-hair hygrometer, 67
humidity measurements, 64-67, **64**, **65**, **66**
hydraulics, 110-112, **110**, **111**

I

incandescent vs. fluorescent lights, 22

instrument flight first, 8-11, **8**, **9**, **10**
International Date Line, 32-33, **32**, **33**, **34**
interoceptor sense organs, 98

J

Jannus, Anthony H., early parachutist/pilot, 2, 7

K

Kelsey, Ben, copilot on first instrument landing, 11
Kittinger, Joseph, free-fall parachute jump, 7

L

levers as early machines, 13-14, **14**
Libyan desert, hottest place on Earth, 71-74, **71**, **72**
lightning and electrical discharge, 22
lissajous (sound) patterns, 118-119, **118**
longitude, 32-33, **32**, **33**, 37

M

machines, 12-15, **12**, **13**, **14**, **15**
magnetic fields, 106-107, **106**, **107**
Martin, Glenn, early pilot, 7
Maxwell, James Clerk, electromagnetic laws, 101
mechanical engineering, 1
   Archimedes, scientific laws of levers, 13, 15
   basic machines, 12-15, **12**, **13**, **14**, **15**
   Channel Tunnel, 18, 19
   dental drill invented, 19
   drills as simple tools, 16-19, **17**, **18**
   force-multiplying levers, 14
   fulcrum-type levers, 13, **13**, **14**
   levers as early machines, 13-14, **14**
   motion-multiplying levers, 14
   pulleys, 15
   Seikan Undersea Tunnel in Japan, world's deepest, 19
   wheels and axles, 15
Merak, pointer star for Big Dipper, 26, **27**
meridians of longitude, 37
metamorphosis, tadpole-to-frog, 91-93, **92**, **93**
meteorology, 59-82
   absolute zero, 103
   Antarctica, coldest place on Earth, 72-74, **72**, **73**, **74**
   Arica, Chile, driest place on Earth, 77, **78**
   barometric measure of air pressure, 63
   "brickfielder" winds of Australia, 82
   Celsius/Fahrenheit temperatures, 68-70, **68**, **69**
   centigrade temperature scale, 69

Cherrapunji, India, wettest place on Earth, 77, **78**
coldest place on Earth, 72-74, **72**, **73**, **74**
computers in meteorology, 59
Dallol, Ethiopia, hottest average climate on Earth, 73
early Greek studies of meteorology, 63
George V Coast, Antarctica, windiest place on Earth, 80, **81**
greatest temperature change in U.S., 74, **74**
hottest place on Earth, 71-74, **71**, **72**
human-hair hygrometer, 67
humidity measurements, 64-67, **64**, **65**, **66**
hygrometers, 64-67, **64**, **65**, **66**
Libyan desert, hottest place on Earth, 71-74, **71**, **72**
Mount Washington, NH, highest wind gusts on Earth, 80, **81**
pumps vs. air pressure, 63
rain forests, 77
relative humidity, 67
satellites in meteorology, 59-60
Vostok, Antarctica, coldest average climate on Earth, 73
weight of atmosphere, air pressure, 61-63, **61**, **62**
wettest place on Earth, 75-78, **75**, **76**, **77**, **78**
wind chill, 82
windiest place on Earth, 79-82, **79**, **80**, **81**, **82**
windmills, 82
World Meteorological Organization, 60
molecular mixture (diffusion), 55-56, **55**
molecular separation (chromatography), 57-58, **57**, **58**
motion, 101
motion-multiplying levers, 14
Mount Washington, NH, highest wind gusts on Earth, 80, **81**
museums listing, 121-130

N

Newton, Isaac, laws of motion and gravity, 101
North Star, 26-28, **27**, **28**

O

Orion constellation, 26, **28**
oxidation (rust) formation, 49-52, **49**, **50**, **51**
  candle and steel wool, 104-105, **104**, **105**

P

Palomar Observatory, California, 25
parachutes, first jump, 4-7, **4**, **5**, **6**

Pasteur, Louis, sterilization techniques, 83
pasteurization, 83
penicillin, 83
Pheil, A.C., first airline passenger, 2
photographing star tracks, 29-31, **29**, **30**, **31**
photosynthesis in plants, 88
physics, 101-119
  absolute zero, 103
  air pressure and floating objects in water, 110-112, **110**, **111**
  air pressure causes heat, 102-103, **103**
  Bernoulli effect, 113-114, **113**
  dolphin deep-dive records, 112
  Earth as giant magnet, 107
  electromagnetism, 101
  friction of air, 102-103, **103**
  gravity, 101, 108-109, **108**
  gyroscopic rotation, 101
  Isaac Newton, laws of motion and gravity, 101
  James Clerk Maxwell, electromagnetic laws, 101
  magnetic fields, 106-107, **106**, **107**
  motion, 101
  oxidation with candle and steel wool, 104-105, **104**, **105**
  sound (lissajous) patterns, 118-119, **118**
  spontaneous combustion, 105
  Trieste submersible, deep-dive record, 112
  venturi effect, 115-117, **115**, **116**
  Weddell seal deep-dive records, 112
  welding temperatures, 105
plants, seeds, germination, 84-86, **84**, **85**, **86**
Polaris, 26-28, **27**, **28**
Prime Meridian, 35-38, **35**, **36**, **37**, **38**
proprioceptive sense organs, 98-99
pulleys, 15
pumps vs. air pressure, 63

R

radiant (solar) energy, 39-41, **39**, **40**
rain forests, 77
Rankin, William H., parachute jump, 7
recycling aluminum, 48
relative humidity, 67
Rigel star, 26, **28**
ripening process of fruits, 89-90, **89**
root hairs and plant absorption of water/nutrients, 88
Royal Greenwich Observatory, Prime Meridian, 38
rust (oxidation) formation, 49-52, **49**, **50**, **51**

S

salts, 53-54, **53**, **54**
satellites in meteorology, 59-60

seeds
  age vs. germination, 86
  cotyledon, 86
  germination, 84-86, **84**, **85**, **86**
  size varies with species, 86
Seikan Undersea Tunnel in Japan, world's deepest, 19
sense organs, 96-99, **96**, **97**, **98**
smell, 96, **97**, 99
soap, eggs, enzymes, 44-46, **44**, **45**
solar (radiant) energy, 39-41, **39**, **40**
solubles, 53
sound (lissajous) patterns, 118-119, **118**
speed of Earth's rotation, 38
spontaneous combustion, 105
stereoscopic vision, 99
sterilization, 83
submarines, *Trieste* submersible, deep-dive record, 112
symbols used in book, xi

T

tadpole-to-frog metamorphosis, 91-93, **92**, **93**
taste, 96, **97**, 99
telescopes, 25
time zones, 33, **33**, 38
touch, 97, **97**, 99
Trieste submersible, deep-dive record, 112

V

venturi effect, 115-117, **115**, **116**
Vinci, da, Leonardo, 83
vision, 96, **96**, 99
Volta, Alessandro, discovery of electrical force, 24
von Zeppelin, Ferdinand, early airline entrepreneur, 2
Vostok, Antarctica, coldest average climate on Earth, 73

W

water/nutrient absorption by celery, 87-88, **87**
weather (*see* meteorology)
Weddell seal deep-dive records, 112
weight of atmosphere, air pressure, 61-63, **61**, **62**
welding temperatures, 105
wettest place on Earth, 75-78, **75**, **76**, **77**, **78**
wind chill, 82
windiest place on Earth, 79-82, **79**, **80**, **81**, **82**
windmills, 82
World Meteorological Organization, 60
wheels and axles, 15

About the Author

Robert W. Wood has professional experience in aviation science, experimental agriculture, electricity, electronics, and science research. He is the author of more than a dozen physics and science books, as well as several home maintenance books. His work has been featured in major newspapers and magazines, and he has been a guest on radio talk shows around the United States. Some of his books have been translated into other languages, including Turkish.

Bob enjoys the wonders of nature and is always interested in the advances of science and how it affects our lifestyles. Although he has traveled and worked worldwide, he now lives with his wife and family in Arkansas.